▶ Research Theatre, Climate Change, and the Ecocide Project

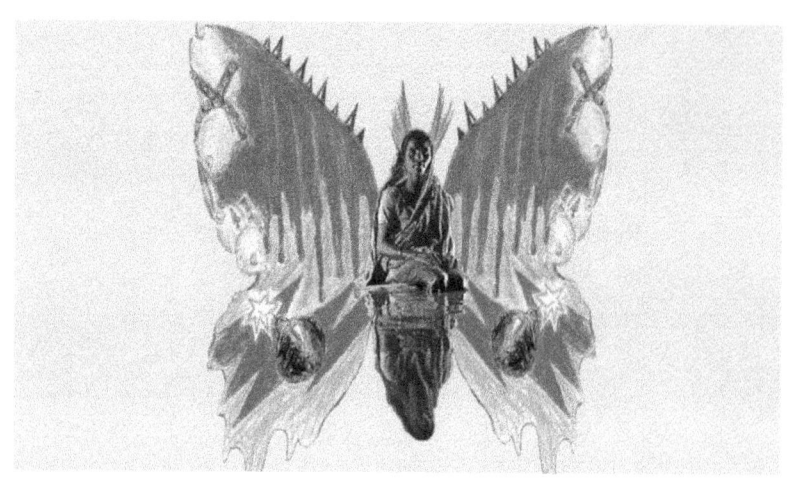

palgrave▶pivot

Research Theatre, Climate Change, and the Ecocide Project: A Casebook

Una Chaudhuri
New York University, USA

and

Shonni Enelow
Fordham University, USA

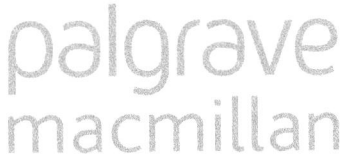

DOI: 10.1057/9781137396624.0001

RESEARCH THEATRE, CLIMATE CHANGE, AND THE ECOCIDE PROJECT
Copyright © Una Chaudhuri and Shonni Enelow, 2014.

All rights reserved.

First published in 2014 by
PALGRAVE MACMILLAN®
in the United States—a division of St. Martin's Press LLC,
175 Fifth Avenue, New York, NY 10010.

Where this book is distributed in the UK, Europe and the rest of the world, this is by Palgrave Macmillan, a division of Macmillan Publishers Limited, registered in England, company number 785998, of Houndmills, Basingstoke, Hampshire RG21 6XS.

Palgrave Macmillan is the global academic imprint of the above companies and has companies and representatives throughout the world.

Palgrave® and Macmillan® are registered trademarks in the United States, the United Kingdom, Europe and other countries.

ISBN: 978-1-137-39663-1 EPUB
ISBN: 978-1-137-39662-4 PDF
ISBN: 978-1-137-39661-7 Hardback

Library of Congress Cataloging-in-Publication Data is available from the Library of Congress.

A catalogue record of the book is available from the British Library.

First edition: 2014

www.palgrave.com/pivot

DOI: 10.1057/9781137396624

We dedicate this book to all our collaborators on the Ecocide Project, especially Fritz Ertl and Josh Hoglund, in celebration of their transgressive imaginations

And to Mohamed Nasheed, who was President of the Maldives from 2008 to 2012, for his imaginative contributions to climate change consciousness.

Contents

List of Figures	vii
Preface	viii
About the Authors	x
1 Research Theatre *Una Chaudhuri*	1
2 Theorizing Ecocide: The Theatre of Eco-Cruelty *Una Chaudhuri and Shonni Enelow*	22
3 A Research Theatre Process: The Ecocide Project *Fritz Ertl*	41
4 Staging *Carla and Lewis* *Ecocide Project collaborators*	62
5 *Carla and Lewis* *Shonni Enelow*	87
Program	117
Bibliography	124
Index	127

List of Figures

1.1	Alex Donis's "Abdullah and Sergeant Adams" (2004). Collection of David Román and Richard Meyer	9
2.1	Amina Crocodile. Photo collage by Sunita Prasad	37
4.1	Amina Butterfly. Photo collage by Sunita Prasad	67
4.2	"Out of the mud come: Crocodiles. Malaria. Rotting wood." Nick Cregor as Landscape (#4745). Photo by Louisa Marie Summer	80
4.3	"What time did you eat dinner?" Nick Cregor "Makes Aquaintance" (#4761). Photo by Louisa Marie Summer	81
4.4	"So long we could feel our cells dividing." Kim Rosen and Meng Ai as Lewis and Carla (#4893). Photo by Louisa Marie Summer	82
4.5	Daniel Squire as Emotional Wall and Nicole Gardner as Mud (#4976). Photo by Louisa Marie Summer	83
4.6	"I'll sleep in mud and die in mud, be born in mud." Nicole Gardner, Daniel Squire, Kim Rosen, and Animated Drawings (#5116). Photo by Louisa Marie Summer	84
4.7	Becoming-Landscape: Nick Cregor, cast members, and animated drawings (#5115). Photo by Louisa Marie Summer	85
4.8	"And I say the crocodile is taking my sister, and I cry the salt of the marsh." Libby King as Elsa-Becoming-Amina (#5249). Photo by Louisa Marie Summer	86
5.1	Polar Bear and Bangladeshi man. Photo collage by Sunita Prasad	109

Preface

Centering on a specific project—encompassing academic research, theatre work-shopping, playwriting, directing, design, production, and theoretical/critical writing—this book offers a practical, theoretical, and critical engagement with the urgent issue of making art in the age of climate change.

The growing scientific and public consensus about the many looming crises following from climate change is matched by an increasing interest, on the part of artists and scholars, to identify creative strategies and practices capable of mounting adequate and appropriate responses to those crises.

The peculiar temporalities involved in climate change pose a challenge not only to our ways of life but also to deeply ingrained disciplinary habits and strongly established frameworks for knowledge production in the arts, humanities, and social sciences. The challenges are particularly acute with regard to the *time scales* these disciplines assume to be relevant, and to their *conceptualizations of human agency*. Collapsing the long-standing distinction between natural history and human history, climate change science proposes a new kind of agency for humans: geological agency, which operates on a scale that not only defies the imagination but also defeats the methods and modes of humanist inquiry. Indeed, the challenge proffered by climate change to conventional ways of thinking is arguably not just an effect of the wreckage it has and will continue to cause, but also one of its *causes*: climate change has advanced to this dangerous verge in part because we haven't known *how to think and feel* about it.

It follows that the imaginative and representational work of making art has an enormous role to play in making this unprecedented crisis visible, audible, and felt. In this book we argue that theatre is a uniquely powerful site for the kind of thinking called for by the crises of climate change.

Using a project we conducted in 2010–11 as a case study, this book theorizes the theatre's potential for new thinking about our time, the Anthropocene, the era when the idea of the human must include a recognition of its shaping role in massive, planetary, geophysical change. The project, called "The Ecocide Project," employed the work process called "Research Theatre," developed by Una Chaudhuri and Fritz Ertl. It used the resources of theatre and performance to locate conceptual and aesthetic principles to represent our time, the Anthropocene, and our present crisis, climate change.

The project unfolded as a series of workshops with two directors (Fritz Ertl and Josh Hoglund), two dramaturges (Una Chaudhuri and Shonni Enelow), and a rotating group of actors. Over the course of two workshops, we explored the set of concepts and figures collected through Chaudhuri's research into what she has termed "theatre of species," including human/non-human "becomings" (Deleuze and Guattari's conceptualization of non-psychological, non-essentialist transformations); the queerness of ecology (described by Timothy Morton and others), with its promiscuous, non-teleological, heterogeneous intimacies; and the theatre of landscape (theorized by Chaudhuri and Elinor Fuchs in *Land/Scape/Theatre*).

The workshops yielded the raw material that eventually became Enelow's one-act play *Carla and Lewis*, which was produced at the Incubator Arts Project in New York in the spring of 2011. (The play was also subsequently presented as a staged reading at Carnegie Mellon University.)

Research Theatre, Climate Change, and the Ecocide Project: A Casebook takes a unique form for an academic publication: it includes both a theoretical investigation of the pressing philosophical and political questions raised by the task of representing climate change, a practical description of the "Research Theatre" process, and the text of *Carla and Lewis*. As such, it is both a handbook for theatre teachers and professionals interested in innovative ways of making theatre, a scholarly analysis of the aesthetic and conceptual strategies needed to do so, and a literary document. We hope this casebook will inspire further reflection on and experiments with theatre as a mode of inquiry.

About the Authors

Una Chaudhuri is Collegiate Professor and Professor of English, Drama, and Environmental Studies at New York University. She is the author of *No Man's Stage: A Semiotic Study of Jean Genet's Plays*, and *Staging Place: The Geography of Modern Drama*, as well as numerous articles on theatre, performance, literature and the environment, and animal studies. She is the editor of *Rachel's Brain and Other Storms*, a book of scripts by performance artist Rachel Rosenthal, and co-editor, with Elinor Fuchs, of the award-winning critical anthology *Land/Scape/Theater*. She was guest editor of a special issue of Yale's *Theater* on "Theater and Ecology," and of a special issue of *TDR: The Journal of Performance Studies* on "Animals and Performance." She is series co-editor of *Critical Performances*, a University of Michigan Press book series that pairs performance artists with theorists/scholars. Her own second volume in that series, *Animal Acts: Performing Species Today*, co-edited with Holly Hughes, will be published later this year.

Shonni Enelow is an assistant professor of English at Fordham University. She received her doctorate in Comparative Literature and Literary Theory at the University of Pennsylvania in 2012. Her recent publications include articles on the poetics of performance documentation (in *Theater*, 2013) and American realist acting (in *Theatre Survey*, 2012). She is completing a monograph on American realism in drama and performance, titled *Method Acting and Its Discontents: On

American Psycho-Drama. Recent theater works include *Carla and Lewis* (Incubator Arts Project, 2011), and a cycle of performance lectures presented in New York at various venues including The Drawing Center, The Poetry Project, and The Invisible Dog Art Center from 2010 to 2013.

palgrave▸pivot

www.palgrave.com/pivot

1
Research Theatre
Una Chaudhuri

Abstract: *This chapter enumerates the principles underlying the practice called Research Theatre, as well as the main ingredients of any given project: an initiating set of research questions, a critical discourse, and a series of improvisatory explorations called "etudes." It then describes three projects that preceded the one that this book is devoted to. The first was a project on globalization and consumer capitalism (The Resistance Project), the second was on the Iraq War (The Queerak Project), and the third was on the new field of Animal Studies (The Animal Project).*

Keywords: Animal Studies, etudes, globalization, improvisation, Iraq, Research Theatre

Chaudhuri, Una and Enelow, Shonni. *Research Theatre, Climate Change, and the Ecocide Project: A Casebook.*
New York: Palgrave Macmillan, 2014.
DOI: 10.1057/9781137396624.0005.

The practice we have come to call "Research Theatre" evolved over almost a decade of collaboration and experimentation. The team of collaborators has always included Una Chaudhuri and Fritz Ertl, and frequently also Steven Drukman, Shonni Enelow, and Josh Hoglund. It has to date encompassed four projects, each addressing a specific but broad area of contemporary critical interest as represented by key theoretical texts and events. The four areas have been globalization, animal studies, war (specifically gender and nationality in relation to war), and climate change—the last of which was the subject of the Ecocide Project, for which the present volume serves as a casebook.

Being the most recent and—we believe—most fully evolved instance of the Research Theatre method, we selected the Ecocide Project for the kind of "360 degree" treatment a casebook represents, but in truth each of the four projects through which we developed this method has generated as rich a range of exercises, discoveries, and creative solutions as were involved in this one. Perhaps the most compelling reason for singling out this project for fuller presentation is its subject—climate change. As the next chapter in particular will argue, the alarming phenomena of climate change—and their implications for our habits of thoughts and modes of life—provide the contemporary conditions for a new sub-genre of theatre practice. We think of this sub-genre as an up-dated ecotheatre, dedicated to putting the vast resources of live, embodied performance at the service of the program of radical reimagination called for by the perilous predicament we find our species—and others—in today. In short, the Ecocide Project produced material that allows us to make a double offering to theatre practice: thus this casebook offers (1) a practical methodology, *Research Theatre* (especially suited to academic and pedagogical settings); and (2) a theatre aesthetic, *Ecotheatre* (particularly attuned to what is arguably the greatest issue of our times).

As the remainder of this book will demonstrate, the version of ecotheatre that emerged from the Ecocide Project was deeply informed by current critical discourses that might loosely be categorized under the rubric of posthumanism, among them queer ecology, vibrant materialism, and nomadic thought.[1] But it was also informed—though less explicitly—by the discourses and discoveries of the three projects that had preceded it. This chapter, then, will offer an overview of those projects, with a view of clarifying the methodology we call Research Theatre but also as a way of introducing certain frameworks of ideas that complement, amplify, and deepen the account of contemporary life, and the proposal for how to engage with it, that was encoded in the Ecocide Project's final product, the play *Carla and Lewis*.

Carla and Lewis is not only a dramatic exploration of global warming; it is very specifically an exploration of global warming *in the age of globalization*, a geo-political phenomenon that has been on our research agenda since our first project, and also featured centrally in our third. I begin, therefore, with a discussion of these two projects.

Our first project, "The Resistance Project," was undertaken in the immediate aftermath of September 11, 2001, the date (and event) from which we might plausibly date something like a dawning of consciousness, among Americans, that their self-image as a nation may not be recognized by people in other parts of the world. The Resistance Project, that is to say, had its start during a period of national confusion and anguished questioning ("Why do they hate us, Mr. Secretary?" a venerable television host asked then-Secretary of State Colin Powell). In New York City in particular, where we live and work, the questions were mixed with sadness and anger, and indeed one of our exercises turned out to involve an emotion-filled visit to what was then called "ground zero" (at that time, before it became routinized, the phrase had a terrifying ring to it, carrying with it many of its chilling associations from the nuclear cold war).

A description and discussion of that exercise may be a useful first introduction to the methodological terrain of Research Theatre, even though it featured a practice that was markedly absent from most of our exercises: silence. We decided to devote part of one work session to a trip downtown during which we all agreed to not speak at all—not to each other, not to anyone else. This "imperative" (as we later called such framing elements of exercises) was a response in part to our sense that *too much* was being said (in the media as well as in our social surround), and that much of what was being said was blocking and distorting our understanding of the event rather than aiding it. Specifically, the amount of talk around us felt like a vociferous denial of a key fact: that this attack was not merely unexpected but *unthinkable*, unfathomable. As such, this was an event that opened what we later came to call—and to seek—"a space of not-knowing."

Such spaces are, we came to realize, emotionally fraught, intellectually repellent, and ... creatively productive. These are the spaces into which a theatre practice that is *conceived as research* could most fruitfully enter, because they are spaces in which questions of what methodologies, affects, or discourses might be appropriate or relevant have not yet been settled. Research Theatre flourishes in these conditions of methodological openness, allowing elements from all of the many channels of performance—space, time, bodies, movement, gesture, thought, emotion, sound, voice, language, music, colors, objects—to be considered for modes of engagement with the questions raised by the conceptually open (or murky) space.

In the case of the Resistance Project and its context in the immediate aftermath of 9/11, the "silence imperative" sought to preserve and expand that space, protecting us from premature decisions and conclusions about what we saw and felt at the site of the attack. It was not until we had returned to our studio—several hours after setting out—that we allowed ourselves to formulate and articulate our experience. When we did, it was clear that the experience had been firmly shaped by the odd interruption of social norms that we had imposed on ourselves. Saved from the deadening requirements of small talk as well as from the daunting prospect of giving adequate expression to powerful emotions, we were able to notice and consider many more elements of our inner and outer realities than is normally possible. Our "research" in this case involved a laboratory-style constraint, a suspension of some features of an event that served to highlight others.

In addition to seeking spaces of not-knowing, Research Theatre often tries to frame an initial question, a "research question" directed at, or prompted by, a critical discourse on a subject of interest to us. This is probably the fundamental difference between Research Theatre and other kinds of exploratory theatrical processes: Research Theatre uses the resources of theatre and performance-making—including such traditional elements as playwriting, rehearsal, design—to delve into *a set of ideas*: a theory, a critical discourse. It identifies key texts and sites of that discourse and engages with them in a variety of ways: reading and discussion, of course (as one would in a seminar), but then also through performance. Performance first takes the form of exercises—called "etudes"—designed by Fritz Ertl and intended to lead us into deeper and more complex consideration of the ideas at hand. These etudes are also intended to "feed" the imagination of the playwright working with us. Once he or she has produced a first draft of a script, performance moves into rehearsal mode, and finally into production design mode. All stages of the work are engaged in a spirit of research, with explicit discussion—involving the whole group—of emerging ideas and questions throughout.

A key principle of Research Theatre might emerge most clearly by way of contrast with another form of devised theatre, documentary theatre. Documentary theatre often involves a great deal of research, and frequently takes the form of interviews with people related to the subject of the play. The goal of this research and these interviews usually tends to be truth and authenticity, hallmarks of the documentarian impulse that are quite at odds with the spirit of Research Theatre. Instead of seeking facts and certainties about a subject, Research Theatre tries to *multiply* the questions, meanings, interpretations, and possibilities evoked by a given discourse. It is this use of theatre as a mode of *further* inquiry, of *extending* investigation that gives us the right, we believe, to use the word "research" as a descriptor of our practice. *Our goal is not to use research to make theatre, but rather to use theatre to do research.* Another way to put this is to say that Research Theatre carries forward the idea expressed in the title of Eric Bentley's classic study, *The Playwright as Thinker*, and asserts that theatre-making itself is a way of thinking.

As mentioned before, the research conducted by and in our projects often begins with a question. In the Resistance Project, the question was "How does one resist (in) America?" It was our way of pointing to the twin challenges of neo-liberalism and consumer capitalism, especially as these are experienced (or, more accurately, *not* experienced, at least not consciously experienced) by young people in America. Specifically, we wanted to explore what happens to the political and personal identities of young people, especially young actors, in the hyper-consumerist logic of late capitalism and globalization, where regional, national, and ethnic identities are appropriated and commodified by multi-national corporations. One of the discoveries of this work that is now part of our process was how the body can be a site of resistance to ideological norms, especially those (like brand-name shopping) that work by rechanneling human desires and drives.

The Resistance Project resulted in a play punningly (and allusively) entitled *Youth in Asia: A Techno-Fantasia (on National Themes)*, written by Steven Drukman and produced by the Department of Drama, NYU, in April 2003. Set in a futuristic commercial-cum-educational institution named "R World"—a cross between Disneyland and a university Theater Department—the play first evoked, and then critiqued, the seductive qualities of consumer culture, its capacity for fulfilling desires that it has itself first aroused. The intersection of consumerism with globalization—and with entertainment (the play used the phrase "weapons of mass distraction" long before it became a meme)—was figured through the "theme-parkization" of ethnic and national identity, by which cultural specificities are reduced to trivial or grotesque—and easily consumable—clichés:

> Sacre Bleu! We lost again! We Belgians will never be as good as the Italians or Argentines at soccer—or football, as it is called in many European nations. [...]
>
> Don't despair. Our indomitable spirit—a mix of Flemish stoicism and French ooh-la-la—is what unifies us! Look at *other* Belgians, well-known personages like Jacques Brel and Hercule Poirot. *They* never gave up!

In the play, this absurd flattening out of anthropological difference was clearly linked to two other phenomena of postmodern, neo-liberal society: the colonization of identity and experience by corporate logics—life defined by brand-names, "logo-life"—and the colonization of lived experience by virtual—mediated and mediatized—modes of representation and communication. In the world of the play, children are selected at birth by corporations and promised life-long sponsorship, in exchange for which parents name and (literally) brand the children for the corporation. Thus there is a character named Nike, who sports a "swoosh" tattoo on her neck. Nike and her co-worker/students (people with names like Avis, Pez, and Domino) in the Asia section of R World are caught up not only in immersive virtual representation—desired places, things, and experiences appear to them almost before they desire them—but also, more troublingly (and presciently), in a system of relentless, wall-to-wall surveillance.

This feature of neo-liberal society, now so well known, led us into the most ideologically challenging discursive territory of the project. As the morbid pun in the play's title suggests, the play entertained the possibility that death is better than life in "R World." We explored one of the more disturbing political ideas that have arisen (on the fringes) of the resistance to neo-liberalism and environmental crisis. We encountered the phenomenon of "suicide environmentalism"—a movement whose slogan is "Save the Planet, Kill Yourself!"—in (among other sites) the tenets of a group that calls itself the Church of Euthanasia and proclaims its devotion "to restoring balance between Humans and the remaining species on Earth. We believe this can only be accomplished by a massive *voluntary* population reduction, which will require a leap in Human consciousness to a new species awareness."[2]

The conclusion our play seemed to reach—that suicide was the only viable mode of resistance in the face of our neo-liberal-totalitarian-consumerist-surveillance society—was markedly at odds with the kind of thinking officially promoted by the "wellness programs" and optimistic activist models that prevail in American academia today. Faculty colleagues expressed dismay and doubt about the direction the project seemed to have taken us. Student audiences, however, were enthusiastic, moved, even grateful. The project's commitment to asking questions (instead of providing answers)—its commitment to being *research* rather than *recommendation*—had cleared a space for thinking and feeling outside official (and officious) norms, and had presented ideas and imagery (for example, the deliberately low-tech recordings of elaborate suicides, on grainy black and white film stock, transmitted as a way to "jam" the slick, color-saturated, and seamless commercial imagery of R World) that truly challenged spectators' modes of living in the so-called real world.

Research Theatre's orientation toward an area of not-knowing yielded even more satisfying results in "The Queerak Project" in 2007, whose explicit premise was the acknowledgement of how little we (average Americans) knew about the culture of the nation we had invaded a few years before. The Iraq war confronted us with another kind of not-knowing from the kind following 9/11: the kind that seems to be deliberately manufactured, for political reasons. This kind of manufactured ignorance is hard to spot, because it is often disguised as information. In the months and years following the invasion of Iraq, the 24-hour news cycle bombarded us with stories and images about Iraq. Certain words and names acquired currency, if not familiarity: Bagram Air Base, the Green Zone, Sadr City. About the actual lives of the actual people there, however, we learned little. Our first impulse—no doubt encouraged by our academic context—was to turn anthropological. Shouldn't we find out as much as we could about the cultural practices and traditional beliefs of the Iraqi people, about their popular culture and local folkways—their festivals, food, fun? However, after quickly reminding ourselves that any valid knowledge about another culture—especially one as different as the Iraqi—would require a much longer period and much more immersive means than we had at our disposal, rather than pursuing a superficial, touristic, and arrogant program of "knowing the other," we set ourselves a different task. The research we needed to do, we decided, had to use the imagination as its main instrument. We wanted to make a connection to this part of the world that would be different—fresher, "queerer," more quizzical—than the ones provided by journalists and talking heads, no matter how embedded or how independent they claimed to be. At first we worked to imagine what it must be like for the young people who have been plunged into that unknown and dangerous reality: our soldiers. We read their blogs and listened to their stories, including those of Sgt. Mkesha Clayton, who spent time with us and generously shared not only her experiences but also her convictions, loyalties, and, above all, her extraordinary spirit. Quite unlike the fact-finding interviews common in devised or documentary theatre, Mkesha's visit did much to strip us of any remaining certitudes or confident political judgments.

Mkesha's accounts of encounters she'd had in the course of her two tours of duty in Iraq also contributed to a second major theme of this project: the role of gender and sexuality in the kinds of violent encounters with other cultures that war often involves. One of the inciting questions posed by our first playwright was whether war could be feminized, and how. The theme of "women in war" gave us a great way to begin to "queer" the Iraq war, that being our term at the time (before the explicitly theorized queer ecologies that were to feed the Ecocide Project) for subverting clichés and conventions and "common knowledge" about a subject. Indeed, the "Queerak Project" had gotten its name from this (admittedly vague) orientation, which unfolded, however, with increasing specificity.

By linking the fact of "not-knowing" Iraq with the embodied and intense kinds of knowing associated with sex, we located a mode of estrangement that, unlike Brecht's alienation effect, was based not on foregrounding the theatrical apparatus but on foregrounding, magnifying, and distorting embodiment. This mode developed during the workshop process and eventually found its way into the play, notably in the form of a prolonged scene in which a group of GIs conduct a "hostile house raid" (a common—and nerve-wracking[3]—tactic to flush out enemies in hiding) into the female protagonist's body, entering through her vagina and crashing around in her uterus.

In addition to an embodied and sexualized "queering" of everything we learned about the war, our workshop process also elaborated what we called "a landscape of catastrophe," another concept that would be greatly elaborated in the Ecocide Project. The landscape of catastrophe that emerged here was both explosively violent and hilariously fake, a product of ignorance and arrogance. It was shaped by several discoveries, including one drawn from deep within the alien culture whose difference we wished to honor. Eschewing ethnographic fantasies of knowing the other—be it the Iraqi other or the soldier other—and reaffirming our desire to use the imagination as a tool of research, we turned to stories, especially myth and folk tales, those imaginative storehouses of a culture's beliefs and values. From among the many Iraqi stories we read, we selected one to focus on, a story of strange couplings and uncanny births, of bizarre captivities and odd outcomes. The story of Husain an Nim Nim, a Tikriti boatman captured, imprisoned, and sexually used by a *s'ilūwa*, a demonic female river-spirit. The *s'ilūwa's* method of incapacitating Husain—she licked his legs till they become spindly and useless—as well as her gift to him—he and his descendants can cure sore eyes with their spittle—all found their way into the theatrical landscape we were constructing, of which the folk tale became one axis.

The second axis was, of course, the war itself, which we first approached through the use of what we call "utopian counterfactuals," imagined—and seductive—"impossibilities." With regard to the Iraq war, one of our utopian counterfactuals was Alex Donis's wonderful painting, "Abdullah and Sergeant Adams" (2004), which shows a GI dancing joyfully with an Arab insurgent.

As the Research Theatre process developed, the idea of impossible dances mixed with the folk-tale's strange account of a disabling love,

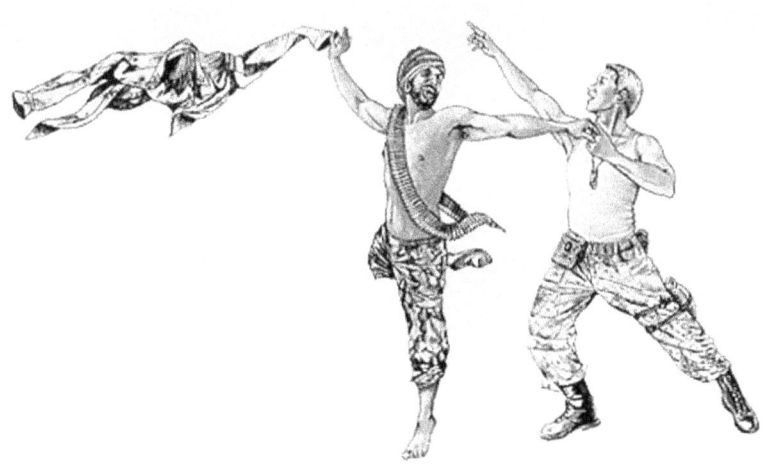

FIGURE 1.1 *Alex Donis's "Abdullah and Sergeant Adams" (2004). Collection of David Román and Richard Meyer*

producing one of our central themes: the role of the *body* in a war supposedly being fought for the hearts and minds of the Iraqi people. This theme was haunted, of course, by the Abu Ghraib images, just as another horrific feature dominating the war news gave us our second major theme: the "weaponization" of children in sectarian conflicts around the world. These themes—broken bodies, sacrificed children, unfathomable wars—often felt too weighty and too serious for our playful ministrations. That was when the spirit of storytelling—as found on soldiers' blogs, in Iraqi folk-tales, and in Mkesha's memories of her tours of duty—came to our rescue. Especially in a war with the specific religious overtones of this one, the question of which stories we share and which ones we don't (or don't know we do) is poignant, even tragic. The biblical story that unites the Abrahamic religions—the story of a perfect garden and a cursed fruit—made a light appearance in our play (there was a garden, and an apple) but it was the surreal spirit of the folktale that produced the landscape we called "Queerak." The folktale also gave us the title of our play: *There Was and There Wasn't*. The phrase is the English translation of the opening formula of Arabic folktales and fairytales (the equivalent of "Once Upon a Time"). Written by Daniel Glenn, and produced by the Department of Drama, NYU, in Spring 2008, it began with a speech by our framing character, THE STORYTELLER:

> *Ahlan wa sahlan*
> Welcome, Americans and others
> When you start a story, you say, "Once upon a time." We do not say this. We say:
> *Kaan yaa maa kaan*
> There was and there wasn't
> or
> *Kaan, maa kaan, ilaa 'an kaan*
> There was, was not, until there was
> And when we are ending a story, we do not say, "And they lived happily ever after." We say:
> *Kaan aku tlaath tuffaaHaat*
> There were three apples
> And then we name who they are for
> One is for the soldier
> One is for the mother of my child
> And one is for the one who will not hear me
> The trick here is we have given three names to the same thing
> It is alright if you do not understand me

Haadhee quSSa
This is a story
NuSha chidhib
Half of it is a lie
And there are many ways to tell it
You will see.

While both the Resistance Project and the Queerak Project used research theatre to investigate two broad subjects (globalization and "war, far away") "The Animal Project" was an explicit engagement with the emerging interdisciplinary academic field of Animal Studies. Our exploration here explored how the conception of "becoming" that underlies philosophers Deleuze and Guattari's enigmatic and challenging notion of "becoming-animal" could be applied to all aspects of theatre: character, space, time, light, sound. The process resulted in the play *Fox Hollow: Or How I Got that Story*, by Steve Drukman, produced and presented by the Playwrights Horizons Theater School.

In this project we developed and discovered many of the techniques of transformation and hybridization that are basic building blocks of a sub-project of Research Theatre that we call "Theatre of Species." A detailed account of this practice was presented in an article published in *Theatre Topics* in March 2006, from which the following account is adapted.

Work on "The Animal Project" began with a focus on four texts: Donna Haraway's little book entitled *The Companion Species Manifesto*, the chapter entitled "becoming-intense, becoming-animal, becoming-imperceptible," in Gilles Deleuze and Felix Guattari's *A Thousand Plateaus*, John Berger's seminal essay "Why Look at Animals?"; and Nobel Prize-winning-novelist J.M. Coetzee's *The Lives of Animals*. Our central research question had to do with the nature of a non-mimetic yet thoroughly embodied and literalistic mode of theatre. What kind of theatricality, we asked, would emerge from the reality claimed by Deleuze and Guattari when they write: "There is a reality to becoming-animal, even though one does not in reality become an animal"?[4]

As work began on the Animal Project, various recent events had drawn different kinds of attention to the place of the animal in contemporary culture. In October 2003, one member of the Vegas animal act team Siegfried and Roy had been mauled, on-stage and before a terrified audience, by one of their performance animals, a White Siberian Tiger named Montecore. At the same time as the infotainment press was issuing hourly bulletins of Roy's condition (and his Christ-like forgiveness of Montecore), across the country, an impoverished man was found to be maintaining a private "zoo," complete with a 350-pound Bengal tiger and a four-foot long cayman, in a tiny fifth-floor apartment in Harlem. A few miles south and a few months later, a different kind of animal habitation drew prolonged media attention: residents of an expensive Fifth Avenue apartment building "evicted" two hawks, Pale Male and his consort Lola, who had taken up residence on a high cornice and were allegedly fouling the area around them. Following a nation-wide outcry and an impassioned vigil outside the building, the nest was restored and the hawks returned. A few blocks from this improbable drama, the Metropolitan Museum of Art's Costume Collection presented "WILD: Fashion Untamed," an exhibition described as "a historical and cross-cultural examination of man's obsession with animalism as expressed through clothing ... [and exploring] the practical, spiritual, psychosexual, and socioeconomic underpinnings of the decorative possibilities of birds and beasts."[5] In an unintended inversion of this project, *W* magazine published a feature on "elephant couture," in which famous designers like Calvin Klein and Dolce e Gabana took up the magazine's challenge of designing outfits for elephants.

All these events helped to situate the project within a kind of "cultural animal unconscious," a web of ideas and images circulating around us, offering clarifications, mystifications, and inspirations. Later, when the text of the play emerged, a central theme (and even plot element) turned out to be just this culture-specific circulation of ideas, linked to the question of knowledge-production as well as to the question how the consciousness of individuals and groups—including non-humans—is altered over time. One of the play's characters is a high school science teacher who is interested in such things as "the Hundredth Monkey" phenomenon, popularized by Ken Keyes's book of that name, which claims (based on an experiment with monkeys on a Japanese island) that when innovative behavior has been adopted by enough members of a group, there occurs an "ideological breakthrough" which allows that new behavior/knowledge to spread throughout the species, without benefit of direct encounter or communication. The implication—that when enough individuals have a good idea and regularly practice it, it will be spontaneously adopted by others without direct contact—in other words, that ideas alone can "change the world"—is as attractive as it is contested (and, in mainstream science, dismissed). Another closely related—and equally controversial—theory that provided us with food for both thought and irony was the theory of "morphic resonance" advanced by the New Age scientist Rupert Sheldrake which explains why (as the title of one of his books puts it) dogs know when their owners are coming home, and other such mysterious behavior.

The notion that there may exist a transpersonal and autonomous kind of consciousness, capable of moving and flowing in unforeseen directions, including across species boundaries, was an elaboration—in the domain of invention and fantasy—of one of the central preoccupations of this project: interspecies communication. This "dream of a common language" (to borrow Donna Haraway's resonant phrase from another context) has haunted the human-animal relation from time immemorial, regularly manifesting itself in culture and science, from Aesop's *Fables* to Koko the signing gorilla. The Animal Project was, to a large extent, an attempt to insert the protocols of theatre and the phenomenology of performance into that age-old (attempted) conversation.

In the course of this exploration, we encountered certain key tropes and traveled some familiar pathways of human-animal interaction, all of which found their ways into the script: scientific experimentation and observation (the aforementioned Hundredth Monkey was performed as a pedagogical "filmstrip," with actors alternating between the roles of scientists and monkeys), magic (the "power-animals" of both traditional and New Age shamanism),[6] and dreams (the literature on dream animals is vast, including classics like Freud's Wolfman and Shakespeare's "translated" Bottom). The major trope for human-animal interaction that emerged in the play, however, was *performance*, represented here by that icon of Western drama, *Hamlet*. Seizing upon Hamlet's characterization of "man" as "the paragon of animals," the play forced an encounter, part-serious and part-ironic, between the heroic humanism of Shakespeare's play and the anarchic animalism of Deleuzian becomings.

From very early on, then, many discussions and improvisations focused on the idea of "becoming-animal," one of animal philosophy's most aesthetically productive of concepts.[7] We began by recognizing first what becoming-animal is *not*: it is not *being*-animal, of course, but it is also not, as many would assume, pretending-, dreaming-, or imitating-animal: "Above all, becoming does not occur in the imagination … Becomings-animal are neither dreams nor phantasies". For our purposes, the most challenging of Deleuzian definitions (or definitions by negation) was the idea that becoming is antithetical to imitation: "We fall into a false alternative if we say that you either imitate or you are. What is real is the becoming itself, the block of becoming, not the supposedly fixed terms through which it passes". Becoming resists metaphor and mimesis. It courts fleeting synecdoches, momentary metonymies, shifting interstices. For actors, it offers an opportunity to indulge and unleash creative impulses without pointing them toward externally (conventionally) settled images. "Animal characteristics can be mythic or scientific," write Deleuze and Guattari, "But we are not interested in characteristics; what interest us are modes of expansion, propagation, occupation, contagion, peopling. ... The wolf is not fundamentally a characteristic or a certain number of characteristics; it is a wolfing".[8]

But the invitation to enter becomings conceived as "wolfings," "lousings," and so on posed another kind of danger: the retreat into private imaginings. We discovered that becoming-animal needed to be carefully distinguished from a sentimental and personalized quest for one's "inner animal," or even one's animal totem or favorite, the one enshrined in habit, personal narrative, and collections. These are, in Deleuze and Guattari tripartite taxonomy, "Oedipal animals, each with its own petty history, 'my' cat, 'my' dog. These animals invite us to regress, draw us into a narcissistic contemplation". By contrast, the animal of becoming-animal arrives from outside. There is nothing familiar, comforting, self-constructing or self-validating about becoming-animal. It is a seizure, a "contagion".[9] This was probably the most original and productive of the ideas we dealt with, challenging us to move beyond the deeply held humanist assumptions (self-determination, intentional, and individualistic questing) upon which most performance training and play-making still rests.

Three other key ideas helped us as we moved our work further into the concept: first, becoming-animal is dynamic and active, continuous and never-ending: a process that never coalesces into a product. Second, the process is an unraveling, a breaking down, a "molecularization," tending toward what the Deleuzian chapter title calls a "becoming-imperceptible." The molecular is opposed, in Deleuze and Guattari, to the "molar," which is the fixed, characterized, constituted, programmed body of "state animals," the second kind in their tripartite taxonomy. The third kind in that taxonomy are the "demonic" animals, which oppose and contest the first two kinds, and are most fully manifested in becoming-animal. Demonic animals disrupt the molar identities of Oedipal and state animals. The notion of the molecular allows Deleuze and Guattari to posit a non-reductive materialism, a reality "that contains no negations or boundaries, but only differences and thresholds".[10] For the purposes of performance, the idea of molecularization and "becoming-imperceptible" functioned as a constant corrective to the pull of mimesis, of inhabiting fixed and recognizable forms and behaviors.

The third idea with which we imbued are our understanding of becoming-animal was that it is a "deterritorialization," a radical dislocation and de-stabilizing of familiar spatial contours and boundaries. In the hyper-semiotic space of theatre, where geography is often destiny and architecture is ideology, deterritorialization is also a potential undoing of the stage and its signifying claims. This idea was taken up most thoroughly by our set and lighting designers, who challenged themselves to produce a space that would both facilitate the actors becomings and perform a spatial becoming of its own.

But any attempt to "script" and "rehearse" a Deleuzian Becoming immediately presents problems and paradoxes so fundamental as to call into question the very legitimacy of the project. How does one choreograph what is defined as an essentially autonomous process? How does one turn an on-going process, without beginning or end, into a "show"? How does one rehearse what must be ever-new, emergent, and spontaneous? And, perhaps most troubling of all: *is it possible for a performed becoming to become real*? Could our journey into becoming finally extend to include the audience? Would our "becoming-theatre" be shared with them, or merely *shown* to them?

This first, almost-immediate encounter with impossibility found its way into the first draft of the script, the one we had in hand when rehearsals began. The problem of staging a becoming-animal manifested itself as an absence, an incompletion: the script called for a play-within-the-play, a production of *Hamlet* put on by some of the characters. In Act I, a rehearsal of this production is shown. However, where, toward the end of Act 2, the actual production is supposed to happen, the first draft of the script presented a blank expanse, with the single word, "Hanimalet."

In this way, an unexpected task was added to the first stage of our work, which we had intended to devote to physical and intellectual animal explorations: we had to "find" *Hanimalet*, we had to figure out how the principle of becoming would transform *Hamlet*. Thus the play-script had incorporated our theoretical investigation; to do our play in the spirit of becoming-animal, we would have to help the playwright discover how to deterritorialize and molecularize this paragon of plays.

In the first sessions of the workshop, Ertl assigned the actors a series of "etudes," designed for them to generate physical language for performing animal-becomings. The original etude had been used as an audition piece: actors were asked to create a brief performance using, as a point of departure, one of the images from Art Shay's wonderful photography book *Animals*. Each performance had to include one transformation from human to animal, and one from animal to human. The transformation etudes showed us how fertile the animal image can be for the theatrical imagination. Vastly different narratives and emotional journeys emerged from the same image. For example, the image of a tiny monkey plastered against a human hand produced a heartbreaking scene of fear and loss for one actor, while for another it led to an explosive encounter with the essential alienation of technology: the little animal in the human had became a cell phone, a tool of urban hyper-activity. Suddenly, as the actor animalized, it sprang terrifyingly to life in his hand, causing him to fling it away in terror. The animal, like the gizmo, is taken for granted, until it forces a recognition of its essential otherness.

Once the workshop began, the transformation assignments got more and more layered, until they included not just one but several transformations to and from "human" to "animal," with various stops in between. The most complex of these explorations came in response to the following etude composed by Fritz Ertl, with the intention of discovering "styles" of animalization as well as modes of moving through and across them, creating a vocabulary and a syntax of transformation, a language of Becoming:

TRANSFORMATION ETUDE
—I am myself, a human animal.
—I transform slowly into an animal.
—I bump back to my human self.
—I transform to a cartoon of the same animal. I speak.
—The real animal rises from within to replace the cartoon.
—A new (second) animal overtakes the original animal from without.
—The original (first) animal emerges to co-exist with the new animal.
—The cartoon animal emerges to co-exist with the new and original animal.
—Your human self emerges to co-exist with the cartoon, original, and new animal.
—Your conglomerate self becomes molecule.

1. Conceive of this etude as a dream that you have. You need not stage the falling asleep or in any way reference the dream, but allow the logic of a dream to inform the experience.
2. Each animal, including your human self, should have a strong primal need (sexual drive, territory, need for comfort, fear, etc). Make that need visceral and manifest.
3. Similarly, each animal senses (sees, hears, smells, etc) differently. Explore the sensory experience of each.

As the actors enacted the transformations, which were not so much versions of becomings as building blocks from which to form them, the human animal and the non-human animal increasingly lent one another behaviors, gestures, physicality. One actor discovered his belt as a tail, catalyzing his transformation into a dog. Another smelled hot dogs and became a pig. From these explorations there emerged a rich vocabulary of gestures and "flows," movements back and forth across human and animal bodies, experiences, worlds.

We began to think of these transformations as belonging to one of two categories: those emerging from "within,"—from intention, memory, gesture, movement—and those taking over from without. The latter became very important in our thinking on the concept of the herd, another central tenet of Deleuzian animal theory. According to "becoming-intense..." animals not only belong to herds, but *contain* herds in themselves. Members of flocks, packs, broods, swarms—their identity is always primarily plural. The image of the herd "overtaking" the individual arose repeatedly in design as well as dramaturgical discussions. Even in its earliest drafts, the play had established the strong role of the herd, albeit with a twist: the herd that was emerging in *Fox Hollow* was a herd of ideas.

The actors' exercises as well as the design team's meetings produced a realization that in order to enact the animal and produce a climactic "becoming," not only our characters and actors, but our play-space itself needed to transform. Despite initial excitement regarding the potential for feats of theatrical design, a transformation using multi-media or another hi-tech apparatus didn't seem quite right: at the moment of transformation, the enactment of "becoming," the focus, we all felt, should be on the living bodies on stage.

Thus etude instructions began to include shifts in perspective, theatricalizations of the animal point of view, with a view of mapping "animal geographies" for our hoped-for "deterritorialized" stage space. The most successful of these were revelatory: not only was the difference of the animal perspective theatricalized but it was theatricalized in a way that embraced rather than ignored the essential unknowability of the animal world. In one particularly memorable experiment, the actor shifted from being a man looking at a dog (and singing "How much is that doggy in the window?") to the dog being looked at, merely by lifting a chair, exchanging the point of view of the man looking down to the animal looking up. Nothing further needed to be shown: the shift was as profound as it was seamless, as clear as it was theatrical. With the perspective etudes, we had stumbled upon a theatrical "deterritorialization" on a micro-level, destabilizing the stage in the most low-tech way possible. Here our ideal, no doubt utopian, was to do the theatrical version of the literary practice that Deleuze and Guattari describe when they say (speaking of Hofmannsthal): "Either stop writing, or write like a rat".[11] Our quest was for a stage adapted not to seeing the animal, but to seeing *as* an animal.

The play that was finally performed that April was not a play any of us could have predicted; although perhaps that was the point. *Fox Hollow, or: How I Got That Story* is about a journalist named Dominique Metropolis who travels to Fox Hollow, a small town in upstate New York, to look into a strange case involving four teenagers. A year before, these four, top students all, had declined offers of admissions to Ivy League universities, declaring that instead of going to college they were going to turn into animals. "It's more honorable to be a donkey," one of them had reportedly opined. Upon arriving in Fox Hollow, "Dom" discovers that what she had assumed to have been a rare, one-time occurrence from a year ago, is happening again, to four teenagers in the current year's senior class. Her investigation of these four leads her to a narrow-minded and over-bearing guidance counselor, a hippie science teacher with dubious intentions and unorthodox teaching methods, her dog, named D.O.G, and a "transplanted transsexual" who calls herself Lina Wertmuller. Lina's "becoming-woman" is one of many becomings the play charts, concluding with the becoming-theatre of the kids' production of Hamlet—"*Hanimalet*."

If the Deleuzian becoming-animal gave the Animal Project its most challenging horizon of ideas, Donna Haraway's little book, *The Companion Species Manifesto*, provided us with a delightfully rich approach to one of the most familiar figures of animal-culture: the pet. Both the director and the playwright were "dog-people," ardently related to the respective dogs in their lives. Inevitably, the play's most literal animal—the only animal character—turned out to be a dog, named D.O.G. for its owner's interest in the fuzzy science theory of "Determined Organic Genesis."

As the dog emerged as the "real" animal of the play, we read a number of other recent texts, such as Paul Auster's *Timbuktu*, Thomas Mann's *A Man and his Dog*, and Susan McHugh's excellent *Dog*, an early volume in Jonathan Burt's beautifully conceived and realized Animal series from Reaktion books. Some of us also attended an east village performance event entitled *Dog Show Party*, in which live television feed from the Westminster Kennel Club show being held in Madison Square Garden was shown accompanied by live "commentary" by performance artists and DJs, dancing by the people in dog suits, and betting on the competing dogs on TV.

Jamie, the actor who played D.O.G., later reported that he had spent long hours at various New York City parks and dog-runs, observing the animals and their owners. He had occasionally followed them on their walks, engaging some of them in conversation and interactions. He had become most fascinated with the dogs' gazes, their way of looking at things, paying attention to some while ignoring others, focusing at times and remaining unfocussed at others. In addition to discovering this new way of seeing, Jamie also discovered—to his great surprise and enjoyment—a new feeling about being seen. He felt his human ego loosen its hold on his experience: "I didn't care if people liked what I was doing. I stopped thinking about how I looked to them. I'd be sitting there drooling on stage and feeling *so* comfortable! It wasn't that I was cutting them off or shutting them down; I just felt I was open to whatever they wanted to feel or think—I had no stake in it."

Many people told members of the cast how much they appreciated the open-ended, even enigmatic quality of our exploration of animals and animality. Many had come to the show expecting to be lectured about vegetarianism and cruelty-free cosmetics. Instead, they found themselves rethinking their views on animals, their relationship to their pets, and the extent to which this culture and all others use animals to think about ourselves, others, and the world. One audience member wrote to us to say: "When I came home from the performance, a man *and* (not *with*) a large, white, shaggy-haired dog emerged from the elevator, and I must say that I regarded them both very differently than I would have two hours before. Now that's the power of theatre."

The Animal Project had, of course, sought to harness the power of theatre to change perceptions and revise assumptions. Beyond that, however, we had also wanted to apply the theatre's analytical powers, to use performance as an instrument of research, a mode of thinking. By submitting the cutting-edge theorizations of animal studies to the test of the embodied imagination, we hoped to extend and deepen our understanding of some of its most challenging concepts.

The research theatre process revealed to us the extent to which the issue of the animal in performance is related to the many emerging theoretical and performative explorations of otherness: how does one investigate a different body/being without interrogating it? How does one estrange without fear? The product of all our readings, all our discussions, all our etudes and compositions, was emphatically not an answer. What the research theatre process yielded—then, as always—was sharper questions, and the desire for more investigation.

Notes

1. See, for example, Catriona Mortimer-Sandilands and Bruce Erickson, eds., *Queer Ecologies: Sex, Nature, Politics, Desire* (Bloomington, IN: Indiana University Press, 2010); Tim Morton, "Guest Column: Queer Ecology," *PMLA* 125 no. 2 (2010); Jane Bennett, *Vibrant Matter* (Durham, NC: Duke University Press, 2010); Gilles Deleuze and Felix Guattari, *A Thousand Plateaus: Capitalism and Schizophrenia* (Minneapolis: University of Minnesota Press, 1987).
2. "Church of Euthanasia," accessed August 20, 2013, http://www.churchofeuthanasia.org/coefaq.html.
3. The terrifying confusion such raids plunge everyone—soldiers and citizens—into is brilliantly dramatized in Rajiv Joseph's Iraq war play, *The Bengal Tiger in the Baghdad Zoo*. In Joseph's play, the confusion is largely caused by the language barrier between the soldiers and citizens, which a translator tries vainly to bridge while everyone is screaming in terror in the dark house. In our play, the confusion was also a result of fear and darkness, but had most to do with the conflation of house and body, a landscape of internalized, embodied, and sexualized violence.
4. Gilles Deleuze and Felix Guattari, "1730: Becoming-Intense, Becoming-Animal, Becoming-Imperceptible…," *A Thousand Plateaus: Capitalism and Schizophrenia*, trans. Brian Massumi (Minneapolis: University of Minnesota Press, 1987), 273.

5 Bolton, Andrew. *WILD: Fashion Untamed*. With contributions by Shannon Bell-Price and Elyssa Da Cruz. New York and New Haven: Metropolitan Museum of Art and Yale UP, 2004.
6 One of the most delightful workshop experiences we included was an afternoon with a New Age shaman, who led us in a process of creative visualization followed by a group discussion in which we shared our imaginal animal experiences in what he called "the spirit world." However, partly in light of readings about traditional shamanism (Bleakley) and largely in the generally ironic spirit of the play, the shaman who appeared in the play-text was a rather comical figure (as were, for that matter, all the spokespeople for various current animal practices/ideologies/obsessions, such as vegetarianism, animal rights activism, etc.).
7 See, for example, Steven Baker, *The Postmodern Animal* (London: Reaktion Books, 2000): 102–104.
8 Deleuze and Guattari, "1730: Becoming-Intense, Becoming-Animal, Becoming-Imperceptible…", 238–239.
9 Ibid., 240, 242.
10 James Urpeth, "Animal Becomings," *Animal Philosophy: Ethics and Identity*, ed. Peter Atterton and Matthew Calarco (London: Continuum, 2004), 103.
11 Deleuze and Guattari, "1730: Becoming-Intense, Becoming-Animal, Becoming-Imperceptible…," 240.

2
Theorizing Ecocide: The Theatre of Eco-Cruelty

Una Chaudhuri and Shonni Enelow

Abstract: *This chapter explores the conceptual challenges of representing climate change and the theoretical underpinnings of the Ecocide Project and* Carla and Lewis. *Chaudhuri and Enelow propose the term "eco-cruelty" to describe the paths of their experiments, which, in contrast to conventional models of ecological theatre, focused on queer ecological intimacy and the lively materiality of the theatre space. Drawing especially from Antonin Artaud's Theatre of Cruelty, as well as Timothy Morton's theories of queer ecology, Lee Edelman's notion of the queer death drive, and Jane Bennett's theory of vital materialism, this chapter locates the Ecocide Project within a broader philosophical conversation and rethinks the terms of traditional ecotheatre beyond place-based practices.*

Keywords: Antoin Artaud, ecotheatre, Jane Bennett, Lee Edelman, new materialism, queer ecology, Theatre of Cruelty

Chaudhuri, Una and Enelow, Shonni. *Research Theatre, Climate Change, and the Ecocide Project: A Casebook.*
New York: Palgrave Macmillan, 2014.
DOI: 10.1057/9781137396624.0006.

It was a bright, sunny day in the postcard-picturesque Maldives, the kind of weather that's made this island nation a holiday paradise, and one of the world's destinations for recreational diving. A different kind of weather, however, and a different kind of diving were involved in the strange performance that occurred on this day in October 2009, a few months before the 2009 UN Climate Change Conference, commonly known as the Copenhagen Summit, or Cop15. With high hopes of making that conference a decisive turning point in the disastrous climate change trajectory that was threatening this, the lowest-lying country in the world—the president and members of his cabinet donned scuba gear and conducted a meeting underwater, seated at tables set down on the ocean floor. Using hand signals and waterproof boards, they signed a petition asking the nations of the world to cut their carbon dioxide emissions.

The Maldives underwater cabinet meeting represents an extreme response to the formidable obstacles that the phenomena of climate change pose to representation of all kinds, including performance. The workshop, the play, and the production that are the collective subject of this volume were responses of a different kind to the same obstacles. While the following chapter will detail the ways in which our workshop used performance to explore certain key issues associated with climate change and its representation, this chapter is devoted to presenting and theorizing the nature of climate change representation, and to explicating the play *Carla and Lewis* as a dramatization of these theorizations.

1 Climate change and the closure of representation

The first thing that makes climate change difficult to represent in art is the maddening fact that climate—unlike weather—can never be directly experienced. As the aggregation of numerous atmospheric and weather phenomena, climate does not manifest itself in any single moment, event, or location. The only way it can be apprehended is through data and modeling—through systems and mediations—all of which have to be processed cognitively and intellectually: have to, in short, be *understood*, rather than *experienced*, phenomenologically and temporally. Another way of putting this is that climate change belongs to a mode of unfolding whose features are inherently resistant not only to representation but even to simple, everyday, embodied observation. This is the mode that Rob Nixon has recently identified as "slow violence," those events whose consequences—though catastrophic—are neither immediate, nor instantaneous, nor spectacular.[1] Slow violence names modes of damage that are attritional, or incremental, and largely unseen, and includes such things as domestic abuse, PTSD, pollution, contamination, deforestation, and the slowly emerging diseases and disabilities that happen in the aftermath of various kinds of disasters, man-made or otherwise. Slow violence is the source of deferred deaths and uncounted casualties; it is the harm that happens after most people have stopped paying attention. A chief characteristic of slow violence is actually its self-camouflage: its signs and effects are typically not thought of as violence at all, because they lack the qualities of explosive immediacy that we typically associate with the idea of violence. Yet, as the phenomena of climate change prove, slow violence has all the destructive power of its more familiar, more temporally bounded version. What it lacks is the eye-catching or page-turning immediacy that would make it readily accessible to vivid representation, either as image, or event, or even sound.

As the harm that happens unnoticed, in the background or in the dark, slow violence is also the ally of destructive forces that want to operate unchecked, forces—like polluting industries, or fossil fuel companies—that are quick to capitalize on the increasingly powerful nexus between representation and spectacle. In a media culture focused on "on-the-spot" coverage of horrendously violent events, including man-made ones like bombings and shootings, and "natural" ones like superstorms and tsunamis, the slowly warming oceans, melting ice-caps, and rising sea-levels of climate change—"disasters that are anonymous and star nobody"—are doomed to media oblivion, unless an inspired climate activist like the (then) Maldives president, Mohamed Nasheed, can come up with a stunt like the underwater cabinet meeting. Such stunts are, of course, the rare exception, and whatever impact they have is soon drowned out (so to speak) by the latest school shootings or celebrity scandals.

A second, more profound way in which the phenomena of climate change resist traditional modes of representation has to do with an intellectual—indeed cognitive—challenge posed by these phenomena. In a series of recent articles, the historian Dipesh Chakrabarty has characterized this situation as one that reveals the limits of traditional modes of inquiry and that requires a reconfiguration of our understanding of what it means to be human. Engaging intellectually with climate change, Chakrabarty argues, isn't a simple matter of establishing new interdisciplinary intersections, new dialogues between the fields of say, history and environmental science, or literary studies and ecology, etc. The peculiar temporalities involved in climate change pose a challenge not only to our ways of life but also to deeply ingrained disciplinary habits and strongly established frameworks for knowledge production in the humanities and social sciences. The challenges are particularly acute with regard to the *time scales* these disciplines assume to be relevant, and to their *conceptualizations of human agency*. Collapsing the long-standing distinction between natural history and human history, climate change science proposes a new kind of agency for humans: geological agency, which operates on a scale that not only defies the imagination but also defeats the methods and modes of humanist inquiry:

> We write of pasts through the mediation of the experience of humans of the past. We can send humans, or even artificial eyes, to outer space, the poles, the top of Mount Everest, to Mars and the Moon and vicariously experience that which is not directly available to us. We can also—through art and fiction—extend our understanding to those who in future may suffer the impact of the geophysical force that is the human. But we cannot ever experience ourselves as a geophysical force—though we now *know* that this is one of the modes of our collective existence.[2]

The idea that climate change redrafts the definition of the human to include, for the first time, a geophysical dimension in it, is reflected in the fact that the scientific community is considering the designation of a new geological period named after humankind. The Anthropocene Age, so christened and proposed by the ecologist Eugene Stoermer and the Nobel Prize-winning atmospheric scientist Paul Crutzen, is the age when the massive influence of human activities and behavior on the ecological systems of the planet has grown to levels that warrant its recognition as a geophysical force. Consensus around this idea has built rapidly, and with it, the focus of climate change denial has shifted: while at first the very existence of climate change was denied, followed a bit later by a denial that climate change was anthropogenic, the denial now centers on questions of what—if anything—can or should be done about it. The debate has moved beyond observable facts, beyond even various theories of their causation, and is now a debate about what might be the right kind of response to those facts, and whether it is too late for any response to work.

Among those who believe it is not too late to do something, there are two clear camps, aligning roughly with the preservationist and the conservationist sides of earlier environmental policy debates. Like the preservationists who were guided by an abiding belief in the wisdom and stability of the "natural world," a group we might call "climate change abolitionists" believe that the only effective response to global warming is to halt—and reverse—all those processes that have led to global warming, chief among them the unprecedented levels of fossil fuel extraction and consumption. Halting the march—and dismantling the edifice—of petro-modernity is seen by them as the only way to save the many species—including ours—that are slated for severe consequences if the trends continue as they are now.

At the opposite extreme, we find those who decry what they see as the technophobia of their opponents, and argue instead for a proactive use of everything at our disposal—including, especially, new technologies—to ensure that the future worlds we make are worlds we actually want to live in. In an essay entitled "Evolve: The Case for Modernization as the Road to Salvation," Michael Shellenberger and Ted Norhaus, who identify themselves as "post-environmentalists," accuse Western "ecological elites" of hypocritically performing "a pseudorejection of modernity, a kind of postmaterialist nihilism," and of preaching a gloomy asceticism while continuing to enjoy the luxurious life-styles of technological modernity.[3] This sentimental "ecotheology," they argue, must give way to a new "secular" view of ecology "consistent with the reality of human creation on Earth [...] a worldview that sees technology as humane and sacred rather than inhumane and profane." In another article in the same volume, entitled "Love your Monsters," Bruno Latour suggests that the way forward will require us to *"modernize modernization,"* and see "human development as neither liberation from Nature nor as a fall from it, but rather as a process of becoming ever-more attached to, and intimate with, a panoply of non-human natures."[4]

Latour's account of human development as growing intimacy with the non-human resonates with several theoretical developments—including "queer ecology" and "vibrant materialism"—that proved extremely generative for our work in the Ecocide Project. Before discussing the ways these ideas emerged and got elaborated in *Carla and Lewis*, we need to acknowledge one final—perhaps the most daunting—cause of the closure of representation we face in climate change. This is the fact that—just as climate is the aggregate of many atmospheric and weather systems—climate change is the aggregate result of countless dispersed behaviors and practices. The "Anthro" in "Anthropocene" or "anthropogenic" is never singular: it is a collective, but not of any kind that we have been used to thinking our individual selves in relationship to: not a collective like a family, a tribe, a nation, etc., but a collective that is in some fundamental and scary way in excess of each of us and all of us, a collective that acts outside our control.

This collective, excessive identity was the first site of resonance between our explorations and the visionary theatre expounded by Antonin Artaud. The kind of surplus of affect and agency involved in reconsidering the human as a geophysical force recalls Artaud's search for a theatre that would surpass individual psychologies and biographies and address itself to the physical organism of each—*and every*—spectator. Culture, Artaud insists, is inseparable from biological functioning: see his celebrated call "to extract, from what is called culture, ideas whose compelling force is identical with that of hunger," as well as his assertion that "we must insist upon the idea of culture-in-action, of culture growing within us like a new organ, a sort of second breath."[5] While Artaud was not, of course, thinking of ecology (much less climate change) in his search for a theatre freed from the cultural freight of Western civilization, his instincts led him to recognize that the theatre is one of the cultural spaces most potentially hospitable to our life as organisms, to our *species life*. A remark of Artaud's could have been the motto for the Ecocide Project and for our search for a theatre of lively materiality: "It is right [says Artaud] that from time to time cataclysms occur which compel us to return to nature, i.e., to rediscover life."[6] This resonance between Artaud's ideas and ours led us to appropriate his notoriously enigmatic and provocative term, "cruelty," and to link it to the climate change context we were seeking to engage. Hence, eco-cruelty.

The new version of "us" suggested by the reclassification of humankind as a geophysical force might be imagined in a variety of ways, and Dipesh Chakrabarty has proposed a particularly lucid formulation. The situation produced by climate change, says Chakrabarty, the need to rethink human agency and the impossibility of doing so within the protocols of the traditional disciplines of the humanities and social sciences, points to a new construction of the figure of the human, one that should displace, or supplement, earlier constructs, including the universalist, sovereign individual of the Enlightenment, as well as the fragmented subject of postmodern and postcolonial theory: "the science of anthropogenic global warming [he writes] has doubled the figure of the human—[now] we have to think of the two figures of the human simultaneously: the human-human and the nonhuman-human."[7] The doubled figure of the human as imagined by Chakrabarty is a useful conceptualization of the materialized and collectivized subjectivity that was at the heart of the project we want to describe here. To make this doubled human a character in our play led us through several contemporary accounts of ecology; it also required us to recognize that certain well-established understandings of ecology—as well as certain key theatre practices based on them and used in much recent "ecotheatre"—were deeply problematic and in need of dismantling.

2 Remaking ecotheatre

The Maldives underwater cabinet meeting incorporates several characteristics and principles of what might now be recognized as "traditional," or "established," ecotheatrical practice, best described and theorized by scholars like Wendy Arons, Theresa May, Baz Kershaw, and Downing Cless.[8] The first of these is literalism, which is closely linked to a second: the practice of site-specificity. An important early stage of ecocriticism, the one dominated by a commitment to the local and the regional, emphasized the importance not only of place but of *specific* places, *actual* places. This tendency was closely related to a growing discomfort with the literary practice of treating the non-human world symbolically—as a metaphor or other trope for human emotions and ideas. Instead of wandering lonely as a cloud, the ecocritic wanted to read clouds—and nightingales, and albatrosses, and leaves of grass, and woods, and snowy evenings, and caged birds—as portals to a more-than-human world, capable of producing experience and ideas that could be at least as important as the ones that arise from our immersion in social worlds and subjective ones. This was, in fact, the view one of the authors of this book (Chaudhuri) urged in an early engagement with the idea of ecotheatre: "Theatre ecology, I believe, will call for a turn towards the literal, a programmatic resistance to the use of nature as metaphor."[9] In the realm of performance, the injunction to deal with "nature itself" frequently led to the practice of site-specificity, or at least of outdoor theatre. Going to the park—if not to the forest— felt somehow more "ecological" than staying cooped up in the black box of theatre, even of the version of it that had had one of its walls removed.

The impulse to displace eco-performance from the cultural space of theatre into the supposedly natural space of a park reproduced a discourse that has come, eventually, to be recognized as one of the very sources of our current ecological crises: the sentimental discourse of a romanticized nature, "capital N-nature," constructed as the pristine opposite of culture. The myth of an untouched nature, an Eden from which our species has carelessly banished itself—which provided a strong foundation for the early environmental movement, especially in the U.S.—has by now been quite thoroughly debunked. Although not everyone can go so far as to accept Timothy Morton's proposal that we get rid of the concept of nature altogether and replace it with the concept "ecology"—understood as fully interpenetrated by the cultural activities of all species, including our own—the idea that some places (like national parks and gardens) are more natural than others (like oil fields, or parking lots) is acceptable only if it's presented within a continuum-model rather than a dichotomous one. The development of fields and concepts like "disturbance ecology" recognize the extent to which what we think of as "nature" has been altered by countless factors—not least human interventions of various kinds—while neologisms like "abnatural" seek to open more conceptual spaces for doing justice to the complexity of the phenomena in question.[10] The realization that "culture" *is* (part of) the nature of our species, and its converse, that the non-human world is both shaped by and experienced through elements of this culture (notably language), is not, for contemporary ecocriticsm, a dead end but just the opposite: the emergence of new arena and new set of modalities for ecological and ecocritical practice.

Instead of embracing the kind of literalism that sent so much early ecotheatre into the outdoors, the Ecocide Project sought another kind of literalism, one that would highlight and mobilize some of the basic features of theatre as a medium: the set-apartness of the theatre space and the separation of actors and audience. Instead of literally removing the walls from the theatre space, we wanted to create a theatre that would literalize and materialize the porousness and diversity of the ecological world, its non-holistic, differential ubiquity. Rather than simply refusing the difference between inside and outside and collapsing the two, we wanted to preserve that difference but treat it as a point of departure for a dynamically interpenetrating world in which the matter inside the black-box of theatre is as alive, as lively (or as "vibrant," to invoke Jane Bennett's important theory[11]) as the matter in a forest or a field.

3 Punk butterflies

The emergent discourse of "queer ecology" offered us a powerful direction for our work, by highlighting the performative, anti-essentialist elements in current scientific readings of Darwin. As the editors of the anthology *Queer Ecologies* write, "cutting-edge ecological thinking understands queer desire to be the quintessential life force, since it is precisely queer desire that creates the experimental, co-adaptive, symbiotic, and nonreproductive interspecies couplings that become evolution."[12] Evolution generates a torrent of life forms that diverge or combine in unpredictable ways, thriving or failing to thrive because of myriad factors, observing no hierarchies or orders, and like, queer theory, rejecting the idea of norms with pathological "deviations." In this view, evolution is neither linear, nor progressive, nor purposive. Rather, it is digressive and transgressive. Timothy Morton has described this "queer ecology" as one where boundaries are "blurr[ed] and confound[ed] at practically any level: between species, between the living and the nonliving, between organism and environment."[13] Queer ecology offers an alternative to the environmental tradition based on capital-N Nature and to its synoptic visions and holistic ideals. "Instead of insisting on being part of something bigger," Morton writes, "we should be working with intimacy."[14]

The first challenge, then, was to come up with a kind of queer ecological figure: a "character," or several characters, who would voice and embody an alternative way of interacting with non-human bodies and landscapes. But if we were to achieve a theatre without nostalgia, and without notes of coercive "togetherness," this figure could not be idealized: not a flower child, earth mother, or animal whisperer, but a disturbing, disruptive presence, who genuinely challenged our values. Intimacy with this creature (or creatures) had to be imposing, fraught, uncomfortable.

The characters whose names, Carla and Lewis, gave the play its title are queer in the boundary-destroying, identity-confounding sense of the term: twins who have sex; who are of one mind, or two or ten; who sometimes seem quite human, sometimes not; who sense things contradictorily, simultaneously, immediately, and without consciousness; who imitate, mirror, and double both the human and non-human bodies around them; who transform without ceremony, as a mode of life, as a matter of course. Like the painted, mugging, shrieking stars of Ryan Trecartin's videos, they're anarchic, but media-savvy, feral, but ambitious.[15] They're also obnoxious, unpleasant, and emphatically not cute. (This "not-cuteness" proved to be one of the hardest things to achieve in performance.) Carla and Lewis are conspecific (two bodies of the same species), "punk butterflies," the play calls them, whose patois recalls modernist theatre's iconic conspecific pairings, like Didi and Gogo of *Waiting for Godot*, reshaped into the uncanny doubling of non-human animals that Deleuze and Guattari evoke in their description of the herd. Like all butterflies, they flutter: their dialogue flits quickly back and forth, disorienting sense and meaning. They are not driven by psychology or intention but rather by contradictory, co-adaptive desires and drives. Like their almost-namesakes, Lewis and Clark, they are peripatetic explorers, who aim not to conquer but to uncover, feel out, and somehow map an unknown and undiscovered landscape.

We found the punk-queer affects of Carla and Lewis, with their non-reproductive, anti-futuristic sexuality, through the Nietzschean affect of Lee Edelman's intoxicating theorizations of "no future." Edelman's articulation of the queer as the limit point of futurity—that which represents the failure of the social order that relies on the figure of the Child to legitimize itself and its violence—seemed to us an apt supplement to Morton's queer ecology of intimacy. As an antidote to any cuteness or coziness that might follow from a discourse of intimacy, we found ourselves returning to Edelman's battle cry: "Fuck Laws both with capital ls and with small; fuck the whole network of Symbolic relations and the future that serves as its prop."[16]

Edelman's anti-futurity also challenges the most obvious way of understanding climate change, as a problem of the future, and forces us to reconsider the stakes of a teleological understanding of ecological time. Our play addresses itself not to future generations but to this place, here and now. In *Carla and Lewis*, this here-and-now is both the theatre itself, and the fictional world inside of it: the New York apartment of curator Elsa Turner who invites Carla and Lewis from Berlin to work on a large-scale installation project she's dreamed up to put "a human face" on climate change. Elsa's idea, as she announces in the play's prologue, is to make people care about climate change through the story of a Bangldeshi climate change refugee named Amina, who lost her home and broke her collarbone when a tidal wave came through the wall of her shack. "Americans must hear her voice," Elsa declares, envisioning an installation in which New York gallery-goers would have video conversations—yes, over Skype—with climate change refugees in Bangladesh, while visual artists interpret the conversations in real time on the walls of the gallery. She has found a Bangladeshi artist named Kamna to do the drawings in Bangladesh, and *Carla and Lewis* to do them in New York.

Elsa's project illustrates the representational logic of humanist ecological thinking and some of the pitfalls of traditional eco-theatre: first locating the reality of climate change out there, far away, with "them," and then trying to bring the problem closer in a safe, "creative" way, instead of starting from the awareness of our uncomfortable, nearness to—sometimes genuinely revolting intimacy with—not just the air and the microbes in the air, but also to the metal chair and the chemicals it's made of, not just the endangered animals but also the DNA of insects, and not just the climate change refugees but also the Bangladeshi mud. What Carla and Lewis do to subvert and transform that thinking and that theatre is the subject of our play. For despite Elsa's efforts to bring the problem of climate change "home" to Americans, she fails to recognize that the mud of Bangladesh is here, right here, in her apartment in fact, covering the floor and soaking through the walls. Elsa can't entirely ignore the mud, but she can minimize it: she thinks it's faulty insulation, and she's been "calling and calling the building management" about it, she says, "for years."

If Elsa's project establishes confines for Carla and Lewis to push against, we, the Ecocide collaborative team, gave them the play they wanted to be in—a play of queer landscapes, becomings, and boundary-blurrings—and let them loose. Carla and Lewis infest Elsa's apartment, taking over her space with their weird (but familiar) paraphernalia (spoons, and duct tape, jugs of nail polish) and appetites (milk, potato chips, prescription drugs). From the moment they enter the apartment, they sense the mud and all of its elements, not, initially, as horrific and sickening, but as "delicious," "in between my toes," "like sex." When Elsa shows them a picture of Amina in her destroyed shack in Bangladesh, instead of pity, what they feel is her sensuous intimacy with the squishy, oozy mud. Elsa can't understand why they don't mirror her emotions about Amina and her story: their "queering" of the image of Amina the sentimentalized subaltern disturbs and unsettles her.

The difference between the kind of emotional mirroring Elsa expects, and the uncontainable doubling that she gets, is one of the "not/but" oppositions of the play: not the mirrors of human communion, but the doubles of the herd. Elsa, it turns out, has her own double, a climate change research scientist named Bronwyn who lives in her apartment building and experiences the mud coming in through the walls in the same way as Elsa does, as an unpleasant but mundane annoyance they can do nothing about. The scientist is the curator's counterpart: like Elsa, Bronwyn is highly emotional about the *Daphnia Pulicaria* (water fleas) her lab is trying to save, but sees no fundamental intimacy between her own life and theirs. When Elsa introduces Bronwyn to Carla and Lewis, they pick up on the self-protectiveness inherent in her authoritative pose and shrill self-righteousness, and provoke her with a series of Nietzschean self-declarations:

> **Lewis:** We aren't scared of acid rain.
> **Carla:** We ARE goddamn acid rain!
> **Lewis:** It's coming, we're fucked, and we know it, we know it so well we can taste it. We're not scared of the polar ice caps melting!
> **Carla:** We ARE the fucking polar ice caps melting!
> **Lewis:** We are the hurricanes and the tsunamis and the flash floods and the fires. And we are the dead animals. The dead animals falling dead from the dead trees to the dead forest floor covered with other dead animals!

Carla and Lewis reject the paternalism of the curator and the scientist, implicating their own bodies as both cause and effect of the transformations wrought by climate change. Carla and Lewis, without guilt, shame, or moralism, take on the refashioned world of the Anthropocene. They voice the doubled human.

4 Becoming-landscape

If the content of the play concerned the representational dilemmas climate change poses to common ways of thinking, and attempted to imagine a different way of being-with non-humans, the form of the play emphasized the liveliness of landscapes, the aliveness of matter. The stage directions in the first scene describe a theatrical landscape that moves from the real situation at hand (actors on stage in a theatre) to a mud landscape of Bangladesh, to the city of New York:

> *A theatre.*
> *An actor is onstage.*
> *Pours mud all over the floor.*
> *Out of the mud come:*
> *Crocodiles.*
> *Malaria.*
> *Rotting wood.*
> *Rats, preening like birds.*
> *Dead fish.*
> *Computer parts.*
> *Amina.*
>
> ...
>
> *A subway car that is also a Laundromat.*
> *Rats, preening like birds.*
> *Milk and potato chips.*
> *Perpendicular movement: standing vertically, then folding like a screen.*
> *Trees/illegal immigrants.*

Rather than personify or anthropomorphize the mud, we wanted to make it a *thing*—a thing that, as Jane Bennett reminds us, "looks back." With *Carla and Lewis*, we tried to create a theatre filled with what Bennett calls "Thing-Power": "the curious ability of inanimate things to animate, to act, to produce effects dramatic and subtle."[17] In our play, the mud was more than a material presence: it was a non-human force, an "actant," to use Latour's term (used by Bennett as well). It drove the story and affected the characters in both obvious and unexpected ways. But it was also an assemblage, a "living, throbbing confederation" (Bennett) of other, discrete actants, both human and non-human, human-made and non-human-made—rotting wood, computer parts, malaria, dead fish, rats, and a young woman named Amina. This assemblage is called "the mud" and acts *as* mud, but also, at any minute, has the potential to split into separate parts with their own agencies: "precisely because each member-actant maintains an energetic pulse slightly 'off' from that of the assemblage, an assemblage is never a stolid block but an open-ended collective."[18]

Here was another resonance with Artaud, whose messy, unnerving assemblages in *Spurt of Blood* ("feet, hands, scalps, masks, colonnades, porticos, temples, alembics"[19]) evoke not only the destruction of human culture but also the non-human agency he describes in "The Theatre and Culture" as "the revenge of *things*": "all our ideas about life must be revised in a period when nothing any longer adheres to life; it is this painful cleavage which is responsible for the revenge of *things*."[20] Artaud's interest in *things* has been little-remarked, eclipsed by his more obvious attention to bodies, but we see in Artaud a vital materialist, who understood the human body in the theatre to be part of a collective of many non-human actants: "we are not free. And the sky can still fall on our heads. And the theatre has been created to teach us that after all."[21]

If queer ecology and "no future" inspired our ecocide characters and their affects, César Aira's novella, *An Episode in the Life of a Landscape Painter*, which details an extraordinary queer-ecological "becoming-landscape," gave us the affect of our landscape, and inspired its climactic transformation. The protagonist of Aira's novella, the nineteenth-century German landscape painter Johann Moritz Rugendas, is a follower of the naturalist and philosophical visionary Alexander von Humboldt, and has traveled to Argentina to document what his teacher calls the "physiognomic totality" of landscape, a striking view of the interpenetration of human and non-human nature: "not in the form of isolated features but features systematically interrelated so as to be intuitively grasped: climate, history, customs, economy, race, fauna, flora, rainfall, prevailing winds."[22] During his journey, two bolts of lightning hit him and his horse, who drags him across the plains, pulverizing his head. He survives, but an exposed nerve leaves him with a massively deformed face, a violent twitch, wrenching migraines, and hallucinations. In Deleuze's terms, he is deterritorialized; wrenched from his subject position as European viewer and privileged representer, he gains access to a new—if extremely painful and nightmarish—kind of sight. When a group of local Indians raid the town where he is convalescing, an event that Aira compares to a typhoon, Rugendas's documentary painting transcends Humboldt's technique and progresses, as Aira puts it, "towards unmediated knowledge":

> Humboldt's procedure was, in fact, a system of mediations: physiognomic representation came between the artist and nature. Direct perception was eliminated by definition. And yet, at some point, the mediation had to give way, not so much by breaking down as by building up to the point where it became a world of its own, in whose signs it was possible to apprehend the world itself, in its primal nakedness.[23]

Aira also showed us that this becoming-landscape—this over-taking of the human body in and through representation—would have to be traumatic, a manifestation of eco-cruelty. The landscape of ecocide would replace the mystical, Edenic fantasies of traditional ecology and capital-N nature with something nightmarish, monstrous, terrifying. What struck us, moreover, was not only the intersection of human and non-human forces—the lightning that penetrated Rugendas's body never leaves him—but also that representation remains Rugendas's modality: indeed, his capacity for representation is perfected by his transfiguration.

Rugendas's traumatic becoming-landscape also inspired another thematic element of our play, in which we translated the idea of deterritorialization into the simple desire and anxiety of leaving one's home. "HUMANS! LEAVE YOUR HOMES!" the actors say, periodically. The phrase—which can be interpreted, and delivered, as a command, or a plea, or a warning—is fundamentally an invitation to rethink the concept of "home" in the context of the unprecedented dislocations of climate change. These dislocations extend far beyond anything that could adequately be captured by the sentimental constructions and discourses of home as hearth, as belonging, safety, community, roots, etc. As the play's conclusion dramatizes, home is now defined by blurred boundaries, affective inundations, and radical deterritorializations. "HUMANS, LEAVE YOUR HOMES!" is eco-cruelty's injunction to the play's characters, actors, and spectators: leave your "comfort zones" and enter new contact zones, leave your organizations and become what famously Deleuze and Guattari found in Artaud's texts: "a body without organs."[24]

At the end of the play, Elsa does leave her home, as she undergoes a becoming-landscape that, like Rugendas's, is both traumatic and beautiful. Carla and Lewis don't require this metamorphosis; they belong to a species that is intimately connected to the landscape from the beginning, and relish its complexity as well as register its dis-ease. But finally, sick of Elsa's pious moderation and total ineffectuality, and literally sickened by the mud of her apartment (which has given them malaria), they take her treasured image of Amina the sentimentalized subaltern and turn it into an explosive painting made from "mud and shit and milk," a painting not of a victimized brown woman but of a compendium of all the elements of the landscape: as the mud itself.

Elsa at first rejects this vision of the mud-landscape, but Carla and Lewis force her to see it for what it is: the landscape of her apartment as well as Bangladesh, the doubled, non-human world in which she lives and breathes and moves as one species among many, one actant among

Theorizing Ecocide: The Theatre of Eco-Cruelty 37

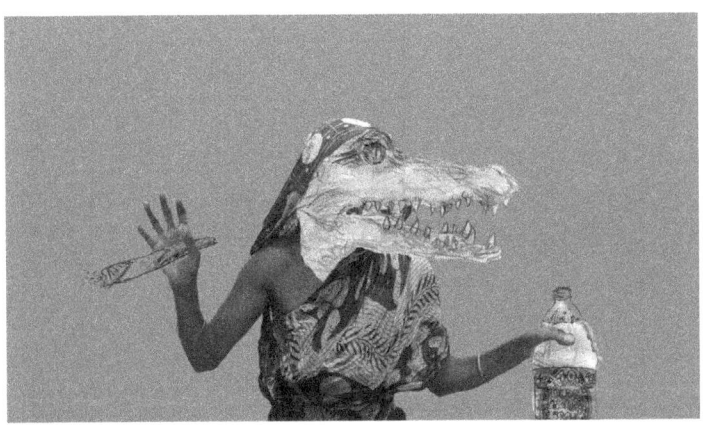

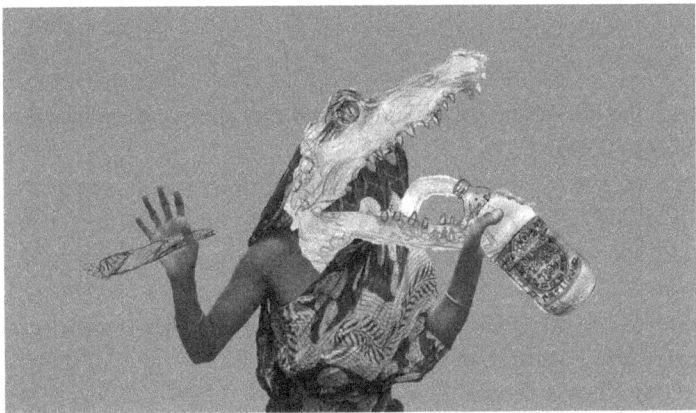

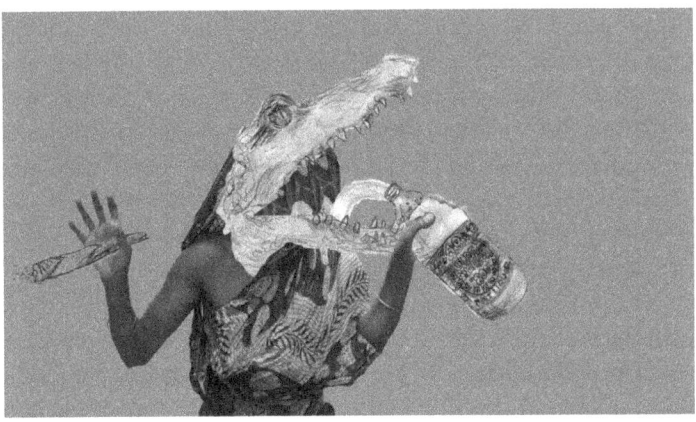

FIGURE 2.1 *Amina Crocodile. Photo collage by Sunita Prasad*

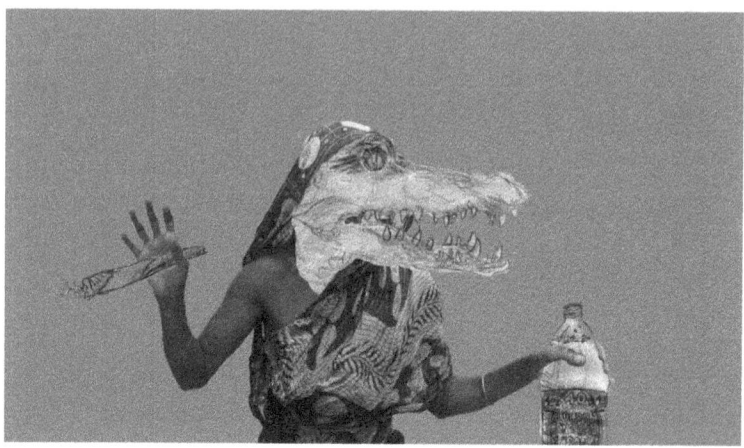

FIGURE 2.1 *Continued*

others, a member of an assemblage she didn't choose, packed with bodies she can't control.

This cataclysm, as Elsa sees the mud, transports her, as well as the play and its audience, into the traumatic becoming-landscape of the flooded Bangladeshi village. The play ends with a long monologue spoken by Elsa-Amina, in which the queer-ecological principles of politicized intimacies reframe the phenomena of climate change as the interpenetration of human and non-human agency.

It starts off as a dream, Elsa-Amina's, who is half-sleeping when the annihilating wave overwhelms her, a dream in which the world is a dark, dead marsh of organic and inorganic, human and non-human matter, a deformed but bewitching world, with "teeth on the branches" of trees, where "the clouds are metal and the sun is a mango," where a crocodile leads her and her sister, the last pair of female doubles in the play, to a palace of colored fabrics, beaks, and feathers, "which are now my sister's feathers." In a watery language of unsubordinated clauses and unfinished thoughts (a bursting, surging language, very different from the tensely fluttering, agitated dialogue of the rest of the play), the actants of the landscape assemblages repeat, refigured: the crocodile, the rotting wood, the metal, Amina's body. When the tidal wave comes, each element splits opens with her chest, a shattering that, paradoxically, dissolves separation: "my chest breaks open and the wall of my house is my head, hard, wet, the mud...which thickens through my neck like

a cake baked in my throat a mud cake." But there is no spiritual unity, no wholeness in this body's dissolution. Here there is nothing but mud: brutal, disgusting, unbearable, full of jagged, broken bodies—and also evolutionary emergence. It is a landscape tense with potentiality: "there is no wall and there is no house and there is no bed and there is no neck and there is no chest and there are no lungs, and there are no bones, the bones are the mud."
The play ends with this clearing.

Notes

1 Rob Nixon, *Slow Violence and the Environmentalism of the Poor* (Cambridge: Harvard University Press, 2011).
2 Dipesh Chakrabarty, "Postcolonial Studies and the Challenge of Climate Change," *New Literary History* vol. 43 no. 1 (Winter 2012): 12.
3 Michael Shellenberger and Ted Nordhaus, "Evolve: The Case for Modernization as the Road to Salvation," in *Love Your Monsters: Postenvironmentalism and the Anthropocene*, ed. Shellenberger and Nordhaus (Breakthrough Institute, 2011). Ebook.
4 Bruno Latour, "Love Your Monsters: Why We Must Care for Our Technologies as We Do for Our Children," in *Love Your Monsters*.
5 Antonin Artaud, *The Theatre and Its Double*, trans. Mary Caroline Richards (New York: Grove Press, 1958): 7–8.
6 Ibid., 10.
7 Chakrabarty, "Climate Change," 11.
8 See Wendy Arons and Theresa J. May, *Readings in Performance and Ecology* (Basingstoke: Palgrave Macmillan, 2012); Downing Cless, *Ecology and Environment in European Drama* (London: Routledge, 2010); and Baz Kershaw, *Theatre Ecology: Environments and Performance Events* (Cambridge: Cambridge University Press, 2007).
9 Una Chaudhuri, "'There Must Be a lot of Fish in That Lake': Towards an Ecological Theatre," *Theater* vol. 25 no. 1 (1994): 23–31.
10 Jesse Oak Taylor, "The Novel as Climate Model: Realism and the Greenhouse Effect in *Bleakhouse*," *Novel: A Forum on Fiction* vol. 46 no. 1 (2013).
11 Jane Bennett, *Vibrant Matter* (Durham: Duke University Press, 2010).
12 Catriona Mortimer-Sandilands and Bruce Erickson, *Queer Ecologies: Sex, Nature, Politics, Desire* (Bloomington: Indiana University Press, 2010), 39.
13 Timothy Morton, "Guest Column: Queer Ecology," *PMLA* vol. 125 no. 2 (2010): 275.
14 Ibid., 278.

15 See the videos of Ryan Trecartin, ubu.com/film/trecartin.html, and Wayne Koestenbaum, "Situation Hacker: Wayne Koestenbaum on the Art of Ryan Trecartin" (*ArtForum International*, June 22, 2009), in which he describes Trecartin's art in terms evocative for *Carla and Lewis*: "I want to ride Trecartin's flying saucers, but I'm also afraid of them. His cosmos, not a tranquilizer, presents a terror-spiked forecast. Of apocalypse-as-party. Of psychological evisceration as spiritual exuberance. Of 'being-at-home' as whirling-dervish danceteria."
16 Lee Edelman, *No Future: Queer Theory and the Death Drive* (Durham: Duke University Press, 2004): 29.
17 Bennett, *Vibrant Matter*, 6.
18 Ibid., 24.
19 Antonin Artaud, *The Spurt of Blood*, trans. Ruby Cohn, in *Theater of the Avant-Garde 1890–1950*, ed. Bert Cardullo and Robert Knopf (New Haven: Yale University Press, 2001): 379.
20 Artaud, *The Theatre and Its Double*, 8–9.
21 Ibid., 79.
22 César Aira, *An Episode in the Life of a Landscape Painter*, trans. Chris Andrews (New York: New Directions, 2006): 6.
23 Ibid., 78.
24 Artaud used the phrase "body without organs" in his radio play, *To Have Done with the Judgment of God*; Deleuze and Guattari explore its implications extensively in *A Thousand Plateaus* in their chapter entitled "November 28, 1947: How Do You Make Yourself a Body without Organs?" after the date that Artaud recorded his radio broadcast. See Artaud, *Selected Writings*, ed. Susan Sontag (Berkeley: University of California Press, 1988), 555–574, and Deleuze and Guattari, *A Thousand Plateaus: Capitalism and Schizophrenia*, 149–166.

3
A Research Theatre Process: The Ecocide Project

Fritz Ertl

Abstract: *This chapter describes, in detail, the workshop activities associated with the Ecocide Project, an example of Research Theatre. It includes discussion of the main questions we were asking in that project, the primary critical works we were using, and the improvisational strategies we used to explore our questions. All three of those elements taken together—critical questioning, readings, and structured improvisations—constitute the practice we call Research Theatre.*

Keywords: becoming, etudes, improvised compositions, landscape theatre, queering, Research Theatre

Chaudhuri, Una and Enelow, Shonni. *Research Theatre, Climate Change, and the Ecocide Project: A Casebook.* New York: Palgrave Macmillan, 2014.
DOI: 10.1057/9781137396624.0007.

Research Theatre begins with a question, or series of questions, about the world we find ourselves living in, and uses critical theory to explore that question creatively *as theatre*. The purpose of research theatre is not to prove any particular theory, but to use theory as one set of tools, among many, in our own search for answers. As such, while critical discourse is introduced into the exploration, so too are paintings and photographs, YouTube videos, news stories, websites, and blogs. This chapter presents a description of the Ecocide Project as an example of Research Theatre, and includes discussion of the main questions we were asking in that project, the primary critical works we were using, and the improvisational strategies we used to explore our questions. (All three of those elements taken together—critical questioning, readings, and structured improvisations—constitute the practice we call Research Theatre.)

Influences, fundamental principle, main methods

Every Research Theatre project has involved working with a writer. In this way, perhaps the most important influence for us has been London-based theatre company Joint Stock, founded in 1974 by (among others) Max Stafford-Clark, whose workshop method led to many of Caryl Churchill's plays in the '70s, '80s, and '90s. Like Joint Stock, Research Theatre uses a variety of methods to explore our subjects, including research, interviews, the personal experiences of actor-participants, and improvised compositions. Workshops normally last for one or several weeks, after which time the writer retires for a period of time, charged with responding to the workshop in the form of a first draft. The play that emerges is not expected to dramatize characters, situations, or locations explored in the workshop; it may, in fact, use as many or few of these workshop discoveries as the playwright wishes. What is important is that the primary question being asked by the project finds a dramatic form that is compelling to the writer's imagination. In short, the playwright is writing in response to the workshop, and not as a transcriber on behalf of the workshop. The actors participating in the workshop, while being asked to improvise and compose scenes are not being charged as writers. Actors understand that the function of the workshop is to fertilize the imagination of the writer, not generate material for the actual play, and that everything generated in the workshop is grist for the playwright's creative mill. This is an important distinction, and one that, again, aligns

Research Theatre with the likes of Joint Stock more than it does with many devised theatre companies, whose actors actively participate in the task of writing the text.

Every Research Theatre workshop is centered on a series of etudes, or compositional challenges, intended to quickly generate theatrical responses to the questions of the workshop. Each etude consists of a series of "prompts" or "imperatives," and actors are given a short period of time, usually a half-hour or less, to build a composition springboarding from the prompts. After the compositions are presented, everyone—actors, directors, dramaturges, and designers—discuss what has been illuminated by each group's interpretation of the etude. Early in the process, these conversations are usually quite long, as complications of the initiating question are articulated, and ideas contained within supporting discourses are explored; but as the workshop goes on, and everyone begins to understand the theoretical context, conversations are shorter and more likely to focus on performance-oriented issues, such as how the actors are solving the demands of the etudes as performers. For the Ecocide Project, if early conversations focused on articulating the theories of Darwin and Deleuze—and how they could be of use to us in an ecological age—by the end of the workshop we were focused entirely on the best performance strategies for embodying the non-human, for *becoming* landscape, and for manifesting geological and microbial time.

While each project necessitates its own set of prompts, there are a number of procedural prompts that have evolved across the four Research Theatre projects, and have become part of our process. The most basic prompt of every etude has been, and continues to be, the imperative to transform, usually multiple times, and across all boundaries, including gender, class, age, and species. Embodying "the other" is a large part of our process; in this project, as will be seen below, by the end of the workshop period we had developed our transformation skills to the point of embodying not just the non-human other, but the inanimate other as well.

A second fundamental imperative of Research Theatre is to "queer," a destabilizing imperative that can be applied to any human endeavor that is linear, logical, and/or hierarchical. Its goal is to lay bare the hyperlogical and repressive structure of human thought and culture, and reveal instead the chaotic, sexually charged truth beneath the surface. To my mind, queering is related to Brecht's imperative to "estrange," with this main difference: Brecht was estranging cultural norms to expose capitalist constructs; Research Theatre, while not denying the importance of the economic sphere, is more interested in exposing constructs that deny the inherent sexuality of existence, those that elevate human reason at the expense of universal chaos. Another way of saying this: if Brecht begins with Marx, Research Theatre begins with Darwin and Deleuze. How these two theorists figured in the Ecocide Project will be further discussed later in this chapter.

Finally, there are a variety of additional performance-related prompts that help the performers to make their compositions: imperatives to use rhythm, tempo, and architecture; imperatives to explore duration, repetition and revision, etc. Many of these were first articulated by choreographers, and brought to the theatre by the creators of the Viewpoints. Given how common these tools are in contemporary theatre making, I will not describe them here. It is important to note, however, that they have proved to be especially generative for the transformative and ecological work that has dominated Research Theatre projects.

The initiating question of the Ecocide Project

The Ecocide Project evolved over nine months, during which time we conducted three separate workshops: a four-day Summer Workshop in July of 2010; a more in-depth Fall Workshop that met every Tuesday night for eight consecutive weeks in October and November of 2010; and a brief Winter Workshop shortly before beginning rehearsals for *Carla and Lewis* in late January of 2011.

The title of the project (Ecocide) makes it clear that we were interested in an ecological exploration; specifically, our questions addressed the darker, ecologically self-destructive impulses of us as a species. Why have we degraded our planet so drastically? Are we destined to be the species that destroys millions of years of evolution in a geological eye-blink, and if so, why? Furthermore, is it possible to rethink how we interact with the rest of the planet, and to re-direct our ecocidal ways towards another kind of engagement? In short, how can we recognize worst selves, admit what is happening, and take on the ecological challenges of our time as the dominating species that we are?

Questions of content invariably give birth to questions of form, and the most burning attendant question in this case was: is it even possible to create a truly ecological theatre? Western drama has always privileged the human experience. An ecological theatre would have to treat the human experience as just one kind of earthly experience among many. It would have to be a non-human—or at least a non anthropocentric—theatre: a "Species Theatre". This is the theatre we began to uncover.

Exploring the idea of what "Species Theatre" might look like, I entered the following into my personal notes:

> In an era of climate change and ecological catastrophe, Species Theatre seeks to rethink and re-imagine the divide between the human and non-human. What if, rather than radically separating ourselves from animals and plants—allowing a privileged few into our living rooms while we ignore, kill, or slaughter the rest—what if, instead, we let animals and plants deep into our consciousness? What if we see each organism, however different, as an ancestor? As a fellow survivor? As a potential sparring partner, or office mate, collaborating with us to survive the ecological mess we find ourselves in? Species Theatre endeavors to give agency to other species—by embodying them, by telling stories from their point of view, and by detailing their experiences in relation to the human experience.

Several weeks before the first workshop, having committed myself to a multi-species perspective, I wrote the cast a letter outlining the goals of the workshop, as well as introducing the underlying ideas as simply as possible. The opening paragraph read:

> I am an animal. A human animal. I share the planet with 10 million other species. However, I know surprisingly little about those organisms, and I know equally little about myself *as an animal*. In other words, I am ecologically ignorant. To compound matters, theatre, the medium I use most frequently to enlighten myself, offers little in the way of help. With the human subject as its central theme, theatre, unlike the visual arts, seems especially ill suited towards an ecologically oriented theatre. The Ecocide Project proposes to discover a way of making theatre *from an ecological perspective*. We hope to make theatre that awakens human understanding of ourselves *as* animals *in relation to* other animals, plants, and organisms. By doing so, we hope to help change how *Homo sapiens* walk through the world.

Evolution and becoming

If maintaining a multiple-species perspective was one constant of the workshop, a second was the imperative to dramatize a landscape wherein life is in constant flux, everything reacting and evolving in relation to everything else. It is here that the influence of both Darwin and Deleuze is most visible. From the same letter to the actors:

> The world we live in is radically unstable, constantly evolving, and with fluid rather than solid boundaries defining individuals and organisms. In Deleuzean terms, it is a world in which we are in a constant state of *becoming* that which we are not—becoming animal, becoming female, becoming other, etc. In Darwinian terms, it is a world in which everything is constantly evolving in response to everything else; a world that repeats itself over and over again across generations, but with an element of radical revision always at play—what Darwin calls "descent with modification." There's no standing still in this constantly evolving, mutating world, only patterns, patterns, patterns...and broken patterns. It is essential that an ecological theatre dramatize this radical instability, this constant becoming—*with characters who have liquid identities and unfixed sexualities; with plots that are based on patterns that mutate; and with a theatrical space that is as unstable as the San Andreas fault.*

Together, Darwin and Deleuze gave us the means to challenge our current ideological biases regarding the supremacy of human reason. Both insist that the universe is not logical, linear, or purposeful, but rhizomatic and without design—unintelligent, when all is said and done; and in such a universe it follows that humans (with their reason) are not the pinnacle of creation, but just another organism still evolving in relation to others. Really taking on what this might mean for us as a species was of paramount importance to us as we began our workshop. In conversations conceiving the Ecocide workshop, we all agreed: the purpose of the project was not to promote political action, but to explore the possibility of a cultural (r)evolution based on a non-heirarchical view of "creation." And Darwin and Deleuze would be our foundational texts, our bibles.

Queer ecology and intimacy

Not surprisingly for a group interested in "queering," we were drawn to a number of discourses exploring the intersections between eco-criticism and queer theory. Two texts of particular importance to us were "Queer Ecology," by Timothy Morton, and *No Future*, by Lee Edelman.

Morton's essay was helpful in establishing how the sexually queer is entirely in keeping with Darwin's theory of sexual selection; in nature, in fact, there is no norm, only a constantly evolving process of selection, enlivened by equally constant diversions and perversions. Morton also introduced us to the idea of ecological "intimacy," that is, the idea that true ecological awareness begins with seeing all of life as a series of intimate interactions, immediate and unpredictable, between all the species on the planet. Life is a series of bodily interactions with seemingly innumerable other organisms, which involve transgressions of boundaries, both beneficial and harmful to the individuals involved. Expanding our sense of what constitutes intimacy was immensely productive for us, and led down a variety of paths, including the exploration of "unseen intimacies," including the 1000 species of bacteria that reside in the human mouth and bowels, and which we depend on for our health. We also explored "forbidden intimacies," more commonly referred to as zoophilia, and the naturalness of cross-species sex as revealed by the commonness of hybridity in "nature." We also became increasingly interested in the intimacy that can happen in a theatre, and the possibility of being intimate with our audiences by appealing to their animality.

The Edelman text, *No Future*, takes on the myth of the child (and by extension procreation) as highly problematic in a world with 7 billion humans. It also theorizes that we, as a species, need to stop romanticizing the future as something that will always be rosy ("The sun will come up, *tomorrow*" sings Annie); Edelman's cautionary stance jibed with our own impulse to "embrace our monsters"—that is, to consider the less optimistic views of our future as we head into the ecological abyss that lies ahead.

Ecocide summer workshop—July 2010

None of the performers who participated in the first ecocide workshop had ever been in conversation with us about Research Theatre, Species Theatre, or Ecocide. As such, we decided to start slowly, introducing the idea of "becoming-animal", and intending to radiate from there into all the other topics discussed. It should be noted that one week before we began, the performers were given a reading packet containing "Queer, Ecology," by Timothy Morton, an excerpt from *No Future*, by Lee Edelman, and chapters from *A Natural History Of Sex*, by Adrian Forsyth. In addition they were given the following Etude to prepare for the first day:

Ecocide etude 1—Becoming Other

Begin by googling a number of visual artists who complicate the human/non-human intersection. Some names to start with: James Balog (ANIMA); John Isaacs (MONKEY, 1995); Daniel Lee (ORIGIN, 1999); Patricia Piccinini (ALICE, 2006); Deborah Sengl (various); Marina Zurkow (SLURB). There are many more, but this will get you started.

Next, choose one image (or sequence of images, or sculpture, or video clip) and prepare a 3- to 5- minute solo response.

Be sure to include at least one transformation. All the artists I mention are exploring human/non human interactions of one kind or another, including hybridity, evolutionary mutation, and forbidden intimacy. Over the course of your exploration you need to embody the entire range of species contained in the image. Follow the artist's lead and break as many boundaries as you can, especially boundaries of identity and place.

Look for patterns. And ways to break those patterns.

Explore instability.

* * *

The primary prompt for this etude is very simple: to transform at least once. However, the assigned readings and the cited artworks helped lead the actors to explorations that went far beyond simple transformations. The most complex solo involved an actor transforming from woman to dog to man to child, and then combining and recombining various aspects of each until it was unclear which of the four distinct beings the actor was at any given moment. Fixed identity had been replaced by a constantly evolving identity—and hybridity. I had hoped to coax out this kind of complicated "becoming" sometime near the end of the workshop, to have it appear in the room during the first hour was a great gift. Additionally, the solo involved the actor touching her genitals as a dog while looking directly at us, unashamedly, thereby introducing the idea of a kind of animal intimacy on day one.

Also adding to the complexity of the responses were the secondary prompts: to take on "hybridity, mutation, and intimacy." Hybridity and mutation are easily understood in a sci-fi context (usually as a negative consequence of some scientific experiment gone wrong), but both terms take on new meaning when viewed as part of the "normal" process of evolutionary becoming. As mentioned above, intimacy is another concept that we were interested in interrogating. Without negating the simple intimacy of human individuals with each other, we were more interested in the intimacy *between* species, both sexually and non-sexually. In terms of "Species Theatre," is it possible to dramatize how humans engage with other species as intimates? Pets come immediately to mind because we live and share so many quiet moments with them, but what *about* all those species of bacteria that inhabit our mouths and bowels? Where might the drama lurk in the microbial?

The second day of the July workshop was spent further exploring an unstable world filled with becomings and surprising intimacies. We also watched and discussed a number of video works that proved to be extremely generative. The first of these, "The Big Bang," by video artist BLU, was immensely successful in portraying an ever-evolving universe filled with surprises. Equally helpful were several video installations by Marina Zurkow, especially "Slurb," and "The Poster Children", which invite the spectator to dissolve the boundaries, not between species, but between human culture and nature. Zurkow's work is also helpful because it takes up the idea that we are already living in a landscape of catastrophe, one in which human-driven disasters, such as climate change and mass-extinctions, are the new norm.

On the third day of the workshop we engaged in the most ambitious compositional challenge of this first workshop. The idea behind this etude was to create an *event* in our little theatre that was dangerously intimate and queered in every respect. Thematically, the groups were charged with taking a "NO FUTURE" point of view. In terms of location, they were asked to set it in some kind of landscape of catastrophe:

Ecocide etude—No Future Art Project

"*Fuck social order and the Child in whose name we're collectively terrorized; fuck Annie, fuck the waif from Les Mis; fuck the poor, innocent kid on the Net; fuck laws both with capital ls and with small; fuck the whole network of Symbolic relations and the future that serves as its prop.*"*—NO FUTURE/Lee Edelman*

Working in groups of 4, create a theatrical, embodied art installation from a NO FUTURE point of view.

Queer theatrical structure—think of this installation evolving or unfolding rather than telling a story.

Queer theatrical space—think about the relationship between viewer and unfolding.

Queer time and space, species, and gender—all transformations are possible, and all boundaries must be broken.
Queer the child—challenge the image of future and innocence.
Set this queered unfolding in a landscape of catastrophe.
Use 3 objects to help define this landscape.
Use at least one "natural" element—water, grass, a flower, rocks, sand, etc.
Explore repetition, pattern, and the breaking of pattern.
Have one moment of complete silence.
Have one section of choreographed synchronicity.

* * *

With eight people participating in the workshop, we had two groups of four each creating a performed response to the challenge. While we didn't know it at the time, this etude may very well have been the prompt for the play *Carla and Lewis* As it would take form nine months later. Especially surprising and useful was the use of objects and elements from the natural world to define the landscape. These compositions were replete with things like eggs and plants and trash barrels, as well as with natural elements like water. All these images would find their way into *Carla and Lewis*. In fact, it was the discovery of the liveliness to be found in the inanimate, of the vibrancy of objects and of the landscapes that contain them—that was the most compelling discovery of this phase, and that propelled us towards the Fall Workshop.

Ecocide workshop 2—Fall 2010

Species Theatre had always been concerned with all the living organisms on the planet, human, animal vegetable, microbial. In the interim between the Summer and Fall Workshops, it became apparent to us that a truly ecological perspective would need to expand beyond life itself, to include weather, the elements, the hydrosphere, forests, mountains, the geological past, as well as mundane objects, including the detritus of 21st-century life. In short, a truly ecological theatre would be impossible without the landscape wherein life takes place.

The artwork dealing with landscape that was most influential in our thinking was *An Episode in the Life of a Landscape Painter*, by Cesar Aira. Of particular interest to us were two aspects of the novel.

First, the protagonist is a disciple of Alexander Humboldt, who first theorized that a landscape was a conglomeration of everything that had influenced the biome in its historical past. Hence, geology and climate were the primary factors, having come earliest, but human culture in all its forms were equally a part of the "natural" landscape once established. This integration of the human into "nature" was very informative in our exploration. The second draw of the novel for us was the fact that it ends with the protagonist "becoming" the landscape he has been struggling to paint. In our early explorations of becoming, such transformations had always involved the human individual becoming some other being, either human or non-human, the idea that one could become the landscape was a large ecological step forward for us.

To begin the second workshop we fashioned an introductory etude that would require the solo performers to treat landscape as another character capable of transformation. It would act as a bridge to the exploration of Aira that was to follow:

Ecocide etude—Evolution on Steroids

Watch the following 4 video clips: BIG BANG (Blu); SLURB (Zurkow); The POSTER CHILDREN (Zurkow); SISSY BOY SLAP PARTY (Guy Maddin). All are hyper-evolutionary, though in very different ways, with different textures, colors, moods, rhythms, and themes. Choose one and respond within the following imperatives:

Create a composition in which there is one central character,

But this character transforms, into different people, different animals, different genders, etc, on their journey towards something.

Along this journey, the space also transforms, shifting landscapes.

Though the landscape need not shift every time the character shifts.

In one of the landscapes is a second character,

Whom the first character interacts with.

At some point the central character becomes the second character,

And the second character becomes the central character.

All of the above can happen in any order you want and resolve however you wish.

* * *

This etude is far more rigorous than the opening etude for the Summer Workshop, aggressively demanding multiple transformations within a shifting landscape, and resulting in a deconstructed identity. There was no prompt towards any kind of intimacy, but the imperative to do so was by now ingrained in the cast, and many of the pieces were memorable because they engaged the audience in ways that made us feel like fellow animals in the journey of the deconstructed protagonist. Actors had also settled into thematic interests, one exploring zoophilia, another queer animality, while a third had become obsessed with owls and machines.

After a few weeks, it was clear that we were ready to move on to new challenges. If etudes are, in fact, "practice pieces", we were playing the scales of Darwinian evolution and Deleuzian becoming with ever-increasing ease. It was time to mutate in another direction.

From the beginning of the fall workshop, we had wanted to discover how to embody landscape. While the first etude introduces the idea of landscape, the way the etude was framed allowed for the performers to treat landscape like the container for their ever-changing, mostly animal characters. At this point we asked: is it possible to treat landscape itself AS character? To help answer this question, we asked the cast to read the novel by Cesar Aira and assigned the following etude.

Ecocide etude—Becoming Landscape

Choose one of the following 5 quotations from AN EPISODE IN THE LIFE OF A LANDSCAPE PAINTER:

1) CLOUDS—*page 9, second paragraph (beginning with "In a few days"... ending with "sheer optics of superposed heights and depths.")*
2) NIGHTMARES—*pages 14-15 (beginning with "peaks of mica kept watch over their"... ending with "But who would believe it?")*
3) LEAVING THE ANDES—*page 16 (beginning with "eventually it became clear"... ending with "Argentina opened before them."*
4) MENDOZA—*pages 21-23 (beginning with "Meanwhile, what he was capable of painting"... ending with "overtaking them all on their journey towards the truly unknown."*
5) BECOMING LANDSCAPE—*pages 32-36 (beginning with "What happened next bypassed his senses"... ending with "there is no predicting the result.")*

Working in pairs, please stage a response to the quotation.

Don't stage the plot so much as bring the landscape itself to life.

Embody as many of Humboldt's elements as possible: climate, history, customs, economy, race, fauna, flora, rainfall, & prevailing winds.
Explore the intimacy of the above elements as they accumulate and interact with one another.
Explore geological and immediate time thru duration.
Use the architecture of the space, and the lights at our disposal, to create changing perceptions of the landscape.

* * *

This etude served to animate the landscape for us, making it not just the background of life, but the aggregate accumulation of life. Everything is part of the landscape. And everything is, if not alive, at least vibrant, and in a theatrical context can be embodied by a human actor. It was immensely pleasurable to watch the compositions that resulted from this etude, as for the first time animal consciousness (human and otherwise) receded into the background, as actors explored the embodiment of a variety of objects and forces within the landscape, including lightning, molecules in the air, mountain fog, and horse-driven carts.

This etude also expanded the formal imperatives that the actor had to work with. The invoking of time and space forced us to introduce the idea of duration. In the novel, Aira describes horse driven carts travelling across the great pampas plains, which are perfectly flat for hundreds of miles. Moving slowly to begin with, and with nothing in the air to impede visibility, the carts would be visible for hours, even days, seemingly not moving. This evocation of geological time, which we cannot see moving because its progress is measured in eons, is crucial to an ecological theatre. In reaction to this discovery, we also played with the opposite of geological time, namely, microbial time, which moves faster than our own time, and can be witnessed only under a microscope. To understand that time passes in many ways is to understand that life is not limited to the human.

Playing with the author

As stated earlier, Research Theatre has always worked with a writer; as with Joint Stock, and in the words of Max Stafford-Clark, the function of the workshops is "to fertilize the writer's imagination." On previous projects, no author had contributed writing of any kind to the workshops, putting off the writing process entirely until the workshops were over. With the Ecocide Project, however, Shonni was interacting with the workshop as a writer from the beginning. In fact, before the Summer Workshop, she had generated an exploratory text, entitled *McCloud*, that contained some of the same characters that would end up in *Carla and Lewis*, and a full day of the summer workshop was spent devising characters and events inspired by that text. All during the Fall Workshop, Shonni generated "Sketches" and "Stage Directions" in reaction to what the actors were devising; we, the directors and actors, would in turn "play" with these sketches—creating etudes on the spot and improvising accordingly. These "Sketches and Stage Directions" reached their highest level of complexity after our work on Aira, and would form the basis of our work for the remainder of the Fall Workshop. What follows is the Sketch/Stage direction Shonni wrote after our exploration of Aira:

Ecocide Sketches and Stage Directions: A theatre, Heat Island, An infant, Turbulence, and Pure Climate:

1. *Scene: A Theatre. In the middle of the theatre a thing does something. A thing still doing something. A thing I can't see. What thing? If you are, I am. If I am, I can't hold onto the ropes on the sides of the theatre on the sides of the door any longer, my hands are raw from the ropes.*
2. *A heat island: a city. (A city is a heat island, warmer than the surrounding country, both because the infrastructures retain heat—concrete, asphalt, other materials—and because there are so many devices that emit heat in urban areas, including groups of bodies, including bodies having sex, rats for instance who are having sex six feet from any human at any given time—). In the middle of the theatre, a city; a heat island.*
3. *If an infant survives, it is so demoralized that consent is almost a logical outcome.*
   Two structures
   One: A hierarchy. This is clean, it is orderly, and it is obvious. It has the authority of the commonplace. There is no yelling. There is speaking loudly, SUCH AS:
   SAVE IT FOR A RAINY DAY
   SALAD DAYS
   RED-LETTER DAY
   HAVE A FIELD DAY
   MONDAY MORNING QUARTERBACK
   PUT YOUR MONEY ON THE LINE

PUT YOUR MONEY WHERE YOUR MOUTH IS
MONEY TALKS
DO YOU THINK I'M MADE OF MONEY?
I LIT THE BRIDGE
I BUILT THE REGION
YOU'VE HEARD OF THE GENOME
ANT AND TERMITE COLONIES
I WILL WEED OUT THE ALIENS
IT IS A SEXUAL FANTASY
LIE DOWN

Second structure: self-organizing. We could say: horizontal, but anyone who has been in a love relationship knows that horizontality is a temporary peace but never a lasting solution. Horizontality is always in danger. Could be vertical at any moment. We become each other and I take over. You take over and I apologize.

WELL I'LL BE A MONKEY'S UNCLE

4. Turbulence: a positive feedback loop. We amplify small deviations until they become major differences. SUCH AS:

Napolean promoted the canned food industry.

All the boys cut their hair and canned themselves.

All the girls touch each other and spread hair life fairy dust over the bedrock of the Martha Stewart Living catalogs of the world.

The insect lumpen-proletariat rises up and reclaims the contents of the industrial kitchen.

An assemblage on the linoleum: blood, guts, and kitchen utensils.

5. Pure climate: Snow covers, titter-titter, are not as pure as the driven snow. "SNOW ON HER LIPS."

Everyone on stage is Boy or Girl Ophelia drowning.

Everyone on stage is watching Boy or Girl Ophelia drown.

Everyone on stage is water.

Everyone on stage is Boy or Girl Ophelia with a brain disease.

Everyone on stage is fucking Boy or Girl Ophelia with a kitchen utensil.

Everyone on stage is the kitchen utensil.

A line.

An edge.

A ledge.

6. Several conversations taking place at the same time. A street corner, a jungle, a bar, an ice cap, a village, an apartment—everyone on stave has something very important to say to someone else. Once it's been said, it's a

climate zone. Uncontrollably but ineluctably there. You can't take it back. You're fucked. Swim down that river.

* * *

In these two succinct pages, Shonni had managed to capture the essential discoveries of the entire Ecocide workshop process. The text is, in effect, an outline for an ecological event, *set* in a theatre, and containing within its very real walls a pulsing city—"heat island." Furthermore, this city is driven by two structures, one rigidly vertical and organized in accordance with a capitalist ethos, the other a more free and personal horizontal construction, and in the midst of these structures is a child, most likely living in poverty, and probably sexually abused. Also present is turbulence, which in the context of this scenario is sexual selection infused with perversions and a logic all its own, and climate, pure climate, which seems to enfold itself around the perversions of life. In short order she has created a landscape that reflects Humboldt's belief that landscape is everything, both cultural and natural. Furthermore, she challenges the traditional image of landscape as a pristine backdrop by making this landscape an urban one, wherein the only fauna mentioned are humans and rats: it is, in short, a landscape of catastrophe. It is also the landscape we found ourselves living in 2010.

This outline was immensely compelling to us as actors and directors. We were thrilled to have something of our own, and that we could embody in a Species Theatre manner. We had two sessions remaining, and we worked feverishly through a series of etudes designed to explore the life on Heat Island. It is important to note that we changed several ways of working for these final etudes. In the past, etudes were performed either solo, or in groups of two, three, or four. For these final etudes, the cast worked as a single ensemble of eight. Additionally, in the past the groups working on etudes had always prepared beforehand, either outside of rehearsal, or during the first 30 minutes of rehearsal. For these final exercises, the cast was given the etudes, and asked to respond to them on the spot, as a group, with no conversation. The idea was to abandon all our assumptions and expectations about theatrical presentation and representation, and inhabit the performance space as fully and exclusively as possible, allowing the compositions to emerge from our intense presence. After three sessions in the Summer Workshop and six sessions thus far in the Fall Workshop, we were all fully versed in the Ecocide way of working. Everyone understood both the importance of transformation and how many different kinds of transformations were possible; everyone understood how intimacy could manifest, both in relation to others on stage, and in relation to the audience; and all understood how they could use rhythm, tempo, duration, and architecture to help compose their improvisations. The group had become a finely tuned ecocidal ensemble, and now faced the challenge of genuine evolution: something messy and inelegant, with neither plan, nor hoped for outcome:

Ecocide etude—Heat Island Rats

Working as a unified group, allow the following to evolve with no preconceptions or planning:

1. *Specify 5 sources of heat.*
2. *Allow the heat to drive all transformations.*
3. *Everyone must become a rat individually.*
4. *Everyone must become a rat in unison.*
5. *Everyone must become a rat in unison again, but with revision.*
6. *At some point the rats and the humans have sex, 6 feet apart.*

* * *

Ecocide etude—Landscape with 2 Structures

Working as a unified group, allow the following to evolve with no preconceptions or planning:

1. *Create a landscape with 2 structures: A = vertical; B = Horizontal.*
2. *Alternate between the two structures at an increasing rate.*
3. *Allow the two structures to become one; i.e, spin them into butter.*
4. *Within the landscape there is a lot of glass and water.*
5. *Bring the audience into your body.*

* * *

The work on these final etudes, designed to explore the details of Shonni's Heat Island Sketches & Stage directions, led us into virgin territory. The etudes up to this point had encouraged structured chaos; we were now asking for pure chaos. Not performance, but real life *evolving* on stage. We stopped the improvs in the early going, if at any time we felt people were performing. Following these admonitions, there were long periods when nothing would happen, the actors simply existing in the space. But eventually one actor would be caused to do something, which would spark a reaction from someone else in the space, and before long the entire company would be working together in real time, BEING the landscape of Heat Island in a variety of different ways. The first etude was easier, if only because it necessitated a detailed interaction between two distinct species: humans and rats. The second etude, with its imperative to embody two different structures, one vertical, the other horizontal (with glass and water), was far more abstract, and required a longer period of exploration before a real landscape could emerge.

The Fall Workshop had taken place at the Invisible Dog, a big gallery space in Brooklyn. Since we worked at nights, the gallery was closed, and we ranged all over the architecturally evocative space, from a little stage space intended for performances, to a long shadowy hallway, to the low-ceilinged basement filled with objects needed for various art projects. It was heavenly. For the final two sessions, however, the gallery space was being used for a special exhibition, and we had to move our workshop upstairs, into a small space used during the day by sculptors working on projects. In addition to being cramped, the space was very dusty, and was lit by glaring lights. There were also filthy carpets rolled up and laying about everywhere. These final etudes, with their imperative to discover true chaos and allow genuine life to evolve, were accomplished in this horrid room, and by the end of each session everyone was covered in dirt and dust, especially those who found they had no choice but to crawl into the filthy rugs during an evolutionary moment of improvisation.

Exhausted yet exhilarated, we ended the Fall Workshop having found real life *on stage* while exploring Heat Island.

Ecocide workshop 3—January, 2011

Technically, there was a third workshop, in late January 2011, right before we went into rehearsals for *Carla and Lewis*, but in fact, that workshop was more about exploring the theatricality of *Carla and Lewis* than developing the script itself. The opening day etude will serve to illustrate how we were working:

Ecocide etude—Carla and Lewis *Landscape*

> A half page into the play, a stage direction describes an urban landscape comprised of 14 elements:

An arc of pure fire
Mud
A crocodile
Malaria
A subway car that is also a Laundromat.
Rats, preening like birds.
Dead fish
Computer parts
Mothballs
Rotting wood
The military
Perpendicular movement: standing vertically, then folding like a screen.
Trees/illegal immigrants
Carla and Lewis.
Working solo, please embody the entire landscape as best you can using the following imperatives:

1. *Rather than working literally, transforming from one element to the next, think of them as an accumulation of intimate encounters. As such, find ways that these elements affect each other, and change each other, and become each other.*
2. *Use repetition and revision as a way to explore Darwinian descent with modification. One example: you may transform from one element to another, but when you return to the first element, you should be changed by that second element.*

2a. *Include one moment of mutation, which is not brought about by interaction, as is normal descent by modification, but is change out of the blue. I move through the world with blue eyes. I wake up tomorrow with brown ones.*

3. *Include a moment of forbidden intimacy.*
4. *Explore the effect of climate, and climate change, on all the elements.*
5. *Explore duration, contrasting elements that evolve slowly with ones that evolve rapidly.*
6. *Use objects, especially ones that are not what they say they are. For example, if you use a mirror, perhaps it is not a mirror in the landscape, but a car door, or someone's dinner.*
7. *Set your landscape in a theatre. Establish an intimate relationship with your audience.*

* * *

The work generated by this etude was foundational to the production of *Carla and Lewis* that followed, as with it we were able to aggressively explore the embodiment of a multi-faceted landscape. The greatest discovery we made was that a complex landscape required a complex weave of strategies in order to embody and represent that landscape. As such, while the crocodile in Shonni's landscape could at times take the form of an actor having transformed into crocodile, it could also manifest, at others, as a plastic, blow-up crocodile found in a swimming pool. Similarly, malaria could be embodied by green light, or an actor buzzing like a mosquito, with equal effect as regards our appreciation of the landscape. Not surprisingly, a Humboldtian landscape, comprised of everything that has had an impact on it—including geology, climate, the non-human, and the human—required an equally complex theatrical interaction, comprised of equal parts performance, choreography, and spectacle.

Conclusion

In July, at the start of the Summer Workshop, we had wondered whether an ecological theatre was possible, and if so, what it might look like. Six months later we had discovered a host of strategies for creating such a theatre, the most important of which was to allow evolutionary life to *actually* take place on stage, rather than the more usual *representations* of life that we were more familiar with. Evolutionary life is made up of infinite encounters and interactions, and everything that comes into contact with anything else generates moments and acts of intimacy. These acts of intimacy, by which beings and objects and elements profoundly affect each other, is precisely the intimacy we had been searching for. To my mind, the sections of *Carla and Lewis* That most effectively achieved this evolutionary life (namely, the opening third of the play) were the most thrilling. Additionally, we had discovered that landscape is an aggregate accumulation of everything in life, and that human actors could embody this accumulation—and by doing so we *could* engage in stories beyond the simply human.

Carla and Lewis is the story of two Homo sapiens who are adapting to life on earth today by evolving towards butterflies. Like butterflies, they are keenly aware of the living landscape that they find themselves in, much more so than their purely human counterparts, and, like typical mutants, seem especially well adapted to the conditions of an urban landscape of catastrophe; additionally, they have no seeming relationship to the growth-oriented economy of the mostly "vertical" world they live in, yet suffer no guilt regarding their own will to survive. They are, in fact, what we all need to become to survive. The play does not waste time arguing about whether climate change is a fact; nor does it suggest that there is still time to avert the catastrophic events that will attend global warming. Rather, it suggests that an awareness of landscape, as well as a willingness to rethink the mythologies of growth and progress that are driving our globalized world, will help us to better cope with the heavy weather ahead.

4
Staging *Carla and Lewis*

Josh Hoglund, Sunita Prasad, Nick Cregor, Meng Ai, and Shonni Enelow

Abstract: *This chapter includes essays on the Ecocide workshop and the* Carla and Lewis *production process from Josh Hoglund, the co-director, Sunita Prasad, the video designer, and Shonni Enelow, the playwright, and a roundtable discussion with Meng Ai and Nick Cregor, performers. Hoglund's essay describes the guiding ideas behind the directorial process and the way he worked with actors. Prasad's documents her working process, personal history with the material, and decision to use GIF animations for Carla and Lewis's drawings. Meng Ai and Nick Cregor describe the unique challenges of performing the non-human and their discoveries throughout the workshop and production process of how to do so. Shonni Enelow explains her writing process and the play's intertextual resonances with theories of modern and postmodern drama.*

Keywords: becomings, etudes, GIF animations, spectatorship, subaltern

Chaudhuri, Una and Enelow, Shonni. *Research Theatre, Climate Change, and the Ecocide Project: A Casebook.*
New York: Palgrave Macmillan, 2014.
DOI: 10.1057/9781137396624.0008.

"How to work with actors in ecological theatre"

Josh Hoglund, Co-Director

I think of the performance script of *Carla and Lewis* as a provocation to break apart the restrictions embedded in theatrical form. As such, *Carla and Lewis* might be thought of as "becoming-theatre": becomings are living ecologies that molecularize, transpose, and transform how we inhabit and interrogate our bodies and our words. This is clear from the first page of the text. How are we to read and activate the bolded lists, the poetic prompt texts (stage directions) that form and reform the play's landscape? In my experience as co-director of *Carla and Lewis*, this is the crux of the script's provocation. If we can track character and landscape through the radical evolution and becomings the text offers in the unstable, fluid form of these poetic prompt texts, then we, the artists and the audience, may come through the other side of the performance mutated, destabilized, irrevocably changed. An ecological theatre insists that lives are interactions, not systems. By embedding prompts (a subway car that is also a laundromat! Rats preening like birds!) in the narrative structure we force the actors out of a psychological, diagnostic, synoptic space of broad observation. We stop thinking about the whole, and gain an intimacy rooted in the pleasure of the encounter. In this space of appearance the actors can play around notions of character and notions of themselves at any given moment. They can try on different bodies and different forms of consciousness and focus on the beautiful, brutal, small interactions that are produced.

We spoke a lot about intimacy while working with the actors, especially with the actors who played the landscape, Nick and Daniel, who physicalized images from the bolded prompt texts and spoke directly to the audience with alternating degrees of charm and menace. It was important that intimacy not be confused with authenticity or originality: there is nothing authentic about these characters, in the sense that we do not know their origins and they live in an unreliable world. Likewise, intimacy does not have to do with vernacular speech or the revelation of some "true self." Real intimacy can be prying, needling, and uncanny. We found intimacy in making the unfamiliar familiar, by interrupting or "glitching" the audience's experience of the narrative with questions common to our bodies: what did you eat for dinner? Did you walk to the theatre? The urgency and proximity of the questions was unnerving, as in Nick's proclamation: "I WANT TO EAT EVERYTHING YOU EAT!"

In the Ecocide Project workshops at NYU and The Invisible Dog, we explored Deleuze and Guattari's notion of becoming. Through Fritz Ertl's etudes, the actors worked with an ecological notion of becoming that integrated Deleuze with Darwin's idea of descent with modification as a radical, jerky evolution of lively biology. These methods of theatricalizing through mutation, repetition, and revision helped to put some of the theoretical ideas of ecology into the actors' bodies. They were given a lot of agency and there was a real sense of play. But the shift from workshop to rehearsal was sobering. Our fluid, queered sense of becoming that remixed voice, self, and body felt stuck in the mud. The characters that pushed our narrative forward seemed like strangers, and I think it was difficult for the actors and for myself to not immediately look to our more conventional acting tools to integrate the perceived psychologies of the characters with our own.

In an ecological theatre, becoming happens around character, not within character. In other words, becoming/evolving is not a process set out on by any individual or a larger narrative. It is not predetermined; it simply is, in a flash, in resonance with a landscape, an evolving whole. For this reason, it seems to me that becoming is easier for both performers and audience to understand when multiple bodies are in action. It is easier see and feel molecularizations in a group: together there is more potential for repetition, juxtaposition, synchronicity, and transformation. Other bodies in a state of play can help individual actors to get out of their heads and respond physically and intuitively. As we moved from the workshops to rehearsal, the introduction of scripted characters and narrative seemed an obstacle to creating intimate encounters with an audience. The actors and I sometimes fell into a trap of trying to justify character's actions in a way that tied their wants to some larger—and ultimately reductive—objective.

Do dramatic irony and character distance an audience from intimacy, and is that intimacy a prerequisite for becoming? This was the question that arose as we worked on the characterization of Elsa, the most human character in the play. Elsa is entrenched in familiar humanist values; her behavior and belief system are easy for an audience to latch on to, as they use recognizable art world stereotypes. Fritz and I worked with actor Libby King to ground her voice and take a light touch when it came to constructing character. Libby owned Elsa's language and voice in her opening monologue, purposefully eschewing any stylization and lending her own physical quirks to the role. But when Libby took on the poetic becoming-refugee/mud text in the final scene, her language, behavior, and style shifted considerably. We worked a lot on breath in these moments to ground the text in the actor's body. The shift in language from Elsa's speech to the poetic text of Amina freed us up from trying to find a hybrid. It was not a merging of Elsa and the figure of Amina; Libby was not reenacting the experience of a climate change refugee. This was first of all because we could not know what that would be like, and second of all, because Amina is a fiction herself. The monologue is not a possession; it is more like an act of witness, an attempt to bear witness to the catastrophic reality of climate change. Libby rooted the poetry of this section in a sensual encounter of the poetic images that she could fathom: the colors, the feeling of water. We, the audience, gain proximity to the refugee through the sensual, poetic language of the text, alongside the actor and the character. The voice of the climate change refugee is an appendage of the becoming. I understand this sort of molecularization, this multiplication of voice and exchange, as a becoming-theatre. For this brief moment, the play proposes the actor (Libby) and character (Elsa) as collective body and consciousness and writer (Shonni) and refugee (Amina), our unknowable other, as our speaker. These fragments of different human experiences can become collective because of their persistence as distinct, individual components.

For me, at some point *Carla and Lewis* ceases to be the story of those two "punk butterfly" artists. Carla and Lewis are our fantasies of a different way of living; they live and breathe alternative hedonism, they are queer to the core, and they demand a new lateral understanding of nature that allows for the human and non-human in a diverse, violent, intimate ecology. They are utopian, but they are grounded in the catastrophic present. But if they are our fantasies, they are not our models. Our world is Elsa's world and the play's journey is linked to her. It is her final monologue that ruptures the playworld with ours. In this final moment the character of Elsa is fractured. The climate change refugee is given voice and her vivid, sensual language brings the audience into close proximity with the catastrophic. The poetry infects our individual bodies as an audience, as a pack. The audience witnesses a queering of the normative constraints of body, voice, speech, and selfhood. The different "I's"—Elsa, the actor playing her, the playwright, the climate-change refugee Amina, and the individual audience member receiving it—all create a cacophony of misrecognition.

This openness to the failure of completeness, and to surprising, delight-filled, unreasonable difference fostered the diverse and generative play that *Carla and Lewis* requires. We tried to be good scientists by telling stories, creating fictions, and setting up experiments to challenge those fictions. Our lively biology uses this scientific process to examine a world without boundaries. The collective cultivation of intimacy through our becoming-experiments, and our work with transformations and poetic prompt texts, propelled our investigation of our landscape, which included the theatre space, the room we were in and the people in it; the fiction, the thing we say it is; and ecology, a living biology, the thing it refuses not to be.

"Unsolicited, Unfiltered Connections: The Ethos of an Open-Source Network"

Sunita Prasad, Visuals and Video

The most generative idea that came through in this process was the idea that the boundary between Nature and Civilization is a fictional construct. Humans are not, as we often presume, outside agents working upon eco-systems. We are encompassed by eco-systems. The built-environments and detritus that we produce are not holes in or partitions from the eco-system; they are part and parcel of the eco-system. Likewise, other species—plant, animal, protozoa—are part of our built-environments and part of our civilizations, as well. To deconstruct that boundary became a major impetus of the design. It influenced the subjects of the images projected as well as my alterations to that content and how it moved on screen. For example, from the starting point of photojournalistic images of people displaced by flooding in Bangladesh, I created animated pastiche creatures, augmenting the figures in the photographs with features and appendages of other animals.

Some of these were animals that have become symbolic of climate change, such as polar bears and mosquitoes. By erasing the boundary between humans and other creatures within the figure itself, we demonstrated the porosity between humanity and other species.

Another important idea, which the text of the play highlights and which I found unavoidable given the photographs we were working with, was the power dynamic inherent in the class, race, gender, and geography of climate change. And yet, even what happens far away to people less fortunate than those of us in the Global North is not really as distant as we might think—or the distance doesn't make much of a difference when we're talking about the health of a whole and singular planet. Our hero, Elsa, is desperate to use her vocation as a curator to draw attention to the plight of "climate change victims" in Bangladesh. However, her elevation of the figure of a particular female-identified victim becomes a specious form of Othering, and her mission to bring art-viewers from the Global North into conversation with the displaced in the Global South quickly comes to seem exasperating and self-congratulatory. Like the project's dramaturge, Una, I have close ties to the region in the photographs. My family is from the Northeastern state of Bihar in India, separated from Bangladesh by just a narrow strip of West Bengal. During visits as a child, I saw flooding in the city of Patna, where trash clogged the drainage systems. The image of a woman in a sari in knee-deep water is not just a photograph for me.

It's difficult for me to analyze how that background influenced my usage of the photographs, but I am convinced that it did. Even when it was determined that what was asked of me by the text was to deface the photographs, I always felt that these alterations, while attempting to channel base instincts, had to also create a pastiche that challenged the boundary between West/East as much as it challenged the boundary between human and other species. Using punk as an anti-authoritarian

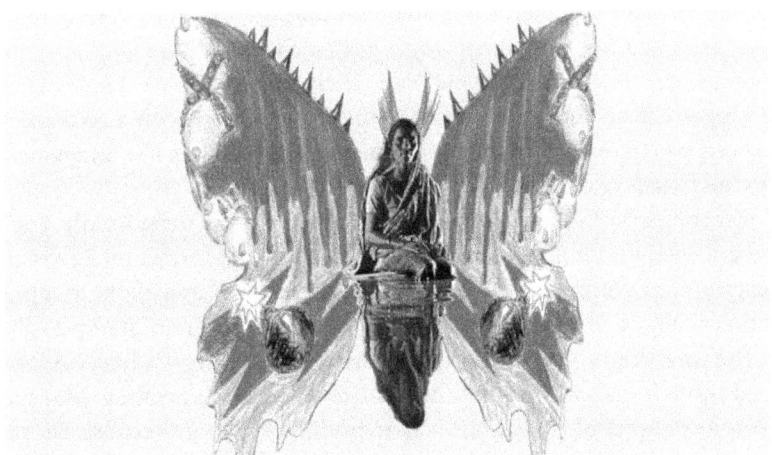

FIGURE 4.1   *Amina Butterfly. Photo collage by Sunita Prasad*

aesthetic readily available in the Western imagination and referenced by Carla and Lewis themselves, I gave the central figure a punk make-over with spikes and tattoos to pivot her victim identification and challenge the authority of Western powers in finding solutions to a crisis which disproportionately affects "Others" in the immediate, but ultimately affects us all.

Collaboration permeated both the content and the construction of this piece. Carla and Lewis are hyper-collaborators who practically share an identity. I think collaboration also holds a certain kind of cachet in the art world, which Carla and Lewis both exploit and are natives of. Most works of theatre rely heavily on collaboration, but I think this production went a step further than most. There were two directors, a writer, and a dramaturge devising work through the improvisations of several performers, cuing the interventions of four designers. I got a little more attention from one of the directors than the other designers did because, at the time, Josh Hoglund and I were roommates. After the first few weeks, we gave up on asking each other whether it was okay to discuss the play while one or the other of us was trying to enjoy a meal, or sitting on the couch with a magazine, or had just arrived home from a freelance job. The creative task at hand permeated our friendship and the walls of our apartment during that period. As you can probably imagine, it was a blessing and a curse.

Fittingly, it created a situation in which it was easy for Josh and I to collaborate on the drawings themselves. We made decisions about the animal collages together. I was adamant that the woman with the polar bear head had to be the one firing a gun. Josh wanted to make sure that enough gore and sexual organs made their way in to have a sufficiently id-tapped effect. And then we sat at the kitchen table with a package of colored pencils and drew pictures together like a couple of momentarily placated kindergarteners.

It is the part of the process that I look back on the most fondly, and I am gratified that it bore some resemblance to what Carla and Lewis are described as doing in the play: locking themselves in a room together and pouring their hearts out through the tips of some colored pencils.

The final scene was always my favorite. Libby King's delivery of that final monologue, when she inhabits the voice of the woman who had been her object of discourse, was incredibly moving. Because we ran our own designs, I was there to see it every night. Of course, maybe in vanity, I can't deny that the scene in which Carla and Lewis's drawings come to life and my animations took center-stage was also gratifying. This was definitely the scene I spent the most time thinking about, in order to produce that animation. After weeks of searching for the visual solution to the text's poetic call-outs in the description of Carla and Lewis's artwork, I found myself returning again and again to animated GIFs (Graphic Interchange Format). Short loops of low frame-rate animation which are a native format of the Internet, GIFs are considered to be among the most accessible networked file formats because they require minimal processing power and network speed. Why did that matter to me in considering this design? Because there seemed to be an analogy available between the porosity and distribution of speciesness and eco-systems embedded in the play and the ethos of an open-source network. Indeed, Elsa is intent on connecting the two sites of her project—Dhaka and New York City—via Skype. She considers this a revolutionary way of producing and distributing socially conscious artwork. I imagined that *Carla and Lewis* would respond to this with all of the unsolicited, unfiltered connections and images that the Internet distributes so readily: GIFs, memes, viral content, and copyright infringement.

Aside from GIF animations' reference to the Internet, and particularly an anti-corporate, open-source philosophy of Internet use, the unfolding of motion in this particular image format produces an immediate and urgent reactivity in its consumption by the audience. I used the lowest possible number of frames to describe motion in the series of animated loops I created as Carla and Lewis's drawings. In just a two-frame loop, punk-butterfly Amina flaps her wings. An angry polar bear woman fires a pistol. Two mosquito-men are disemboweled by a fish. A sari-clad crocodile with a forty and a blunt snaps her jaws.

These colorful, frenetic loops link the mind's interpretation of the images with the body's visceral reaction to their presentation. In an essay entitled "The Affect of Animated GIFs," internet artist and curator Sally McKay describes how "when watching this jerky motion, the viewer's brain becomes actively engaged in the perceptual process, working to fill in the gaps in the action, creating a sense of motion that is never quite seamless, and thus never quite complete as an illusion."[1]

Infinitely delaying the completion of the illusion in Carla and Lewis's artwork distinguished Carla and Lewis's understanding of their subjects from Elsa's. And the affect of the "jerky" GIF animations offered the audience the possibility to experience either (or both) the trances of *Carla and Lewis* and/or the revulsion felt by Elsa.

Carla and Lewis are artists. I spend a lot more time in the visual arts world than in the theatre world, so Carla and Lewis are my peers. I know people who are a lot like them. Certainly who look a lot like them and are very much in their school of direct expression as artists. I am not really in that school. I sometimes wish I were. However, while being vulnerable is a major component of my work, it is done in a very particular, deliberate way. I do a lot of research for most of my projects. I synthesize that material through performances, videos, objects, sounds. Moreover, a lot of my work deals with the politics of gender and my own body, which happens to be marked as South Asian and female. It was difficult for me to suppress the impulse toward political correctness, pick up this really traditional medium of colored pencils, and cover the bodies of South Asian refugees with cartoon blood, tits, dicks, spikes, guns, and booze. I would not have done it at all if I didn't believe that we could achieve a kind of liberation of these images through their re-imagining as embodied agents with desires, emotions, chemical interactions, and an interspecies culture.

"Until We Ourselves Were Elements": Roundtable discussion with Meng Ai and Nick Cregor, performers[2]

SE: Let's talk about the fall workshop and the "heat island" improvisation. What did that feel like, and how did it differ from other performance work you've done?

MENG AI [CARLA]: There was the big challenge of becoming something inanimate. We discovered approaches—abstracting what it is, or de-psychologizing it, or seeing it as a representation of something. That was the challenge: finding approaches we could take, and then finding the most interesting approach. It was exciting when you found the clearest version.

SE: Was there any pattern as to what that clearest version was?

MA: There wasn't one perfect way to do it. There were different modes. I remember for example, dead fish: someone slapped something on themselves, and it was so clear. And then other people embodied the dead fish, heaving for air—physically becoming the fish. Those are both clear, and completely different things. And they served different purposes in the play. We started to discover the different purposes, although they weren't necessarily used to their full potential. I don't think we understood the power of that yet.

NICK CREGOR [LANDSCAPE]: I don't know whether we ever managed to completely bring dead fish alive. I think in general we worked too literally when attempting to bring inanimate, or nonhuman forms to life. On one level, there was the challenge of simply trying to embody or convey an element (those elements which composed the landscape) as accurately as possible. However, there was a second challenge of translating the poetry that existed on the page, painted with those elements, into a performable language that could exist on the stage and in our bodies. Portraying the individual inanimate/nonhuman elements with 100% accuracy still did not insure that we were conveying the landscape with any accuracy at all. This spoke to a common challenge of working with poetic text, and the process of translation from page to stage. One must not translate word by word, but instead have a deep and profound understanding of the larger meaning, and then using the second language with its different and unique tools, reconvey that meaning.

MA: The heat island improvisation was the first time we moved from individual and duo to group work and I was surprised by how well it integrated, by the way we were able to collectively arrive at the same idea. When you're doing it, you feel it becoming this thing. This idea that everyone all of a sudden has.

NC: In improvisation, you rely more on instinct and response. This is what animals do—they rely more on instinct and not on mental processes or psychology. Through that, we started to find things that were less human. We were forced to take in elements that were not just ourselves or our own thoughts—we were forced to take in the space and the architecture, for instance, and what we were bumping up against in that space and time. We weren't doing something narrative—we'd removed the psychology. I was very excited by that. Then we had to figure out the relationship between that kind of improvisation and what we were ultimately going to do in performance.

MA: As actors, we have the habit of setting things. [When we got to the production stage] we set certain things—dead fish became dead fish pretty quickly.

NC: You're talking about the winter workshop, after we had the script?

MA: Yeah. I don't know if that was lazy, or if we were just out of ideas—but it became a shorthand. And then it was less exploratory.

SE: So what was it like to perform or embody the non-human?

MA: I found that really challenging. At times it was an interesting challenge, at other times it was just extremely frustrating. I found some of the [landscape elements] untranslatable. Things like mothballs. It's so specific, so you can't even do the idea of what it is. It's just that thing. Dead fish has a feeling to it.

SE: So when there were more associations or emotions attached it was easier?

MA: Yeah. And also dead fish doesn't have a specific purpose. Whereas mothballs have an actual purpose. And even the purpose is very inanimate.

NC: I think we just nicked the surface. The workshop was just the tip of the iceberg. I personally did a lot of work with de-psychologizing. In order to dehumanize, and be other, and be more interconnected. Instead of ignoring the human that's always been at the center of theatre, how to disperse and fracture that [psychology], because it's just one part of a larger picture? I thought very specifically about animality and psychology, which led me to thinking about mental disabilities. That was the focus of one of [my etudes]; it was one potential launching point for de-psychologizing. As the workshops went on, and we delved into improvisations, we began to discover a lot more about the non-human elements: when they were unspoken, when they were about space, when they were about undercurrents of emotion. Those were exciting, and they started touching on how we were not just human beings in this scenario, how are we connected to each other and to our environment. Understanding in an ecological sense.

SE: What do you think you will able to preserve from the workshop to the production and how do you think it informed your performance?

NC: I remember one day, we were sitting in chairs, and Fritz came up to me and whispered something in my ear. And what he said, or what my memory recreates of what he said, was, "Stop performing." I don't know if "be yourself" was in there, but it was something very simple. And that launched another mode of being, which was this honesty at the root of the performer, which felt very pertinent, perhaps in an analogous way—like what we said in the first day: okay, if this is about ecology and about the spaces we inhabit, we must take into account this theatre space that is our current moment in time to understand who we are in this context, created by these elements. At the end of the day, we are human beings. But to understand that [humanness] at its simplest and most elemental was to speak to something vaster. Interestingly, the landscape's character was an attempt to be this human being at its most simplest and stripped down form. The point wasn't for a performer to recreate a wall or a dead fish, or even to embody it, although in moments these were tools that these performers used, but that at the root, the landscape was Nick and Daniel. And that felt interesting.

MA: It's actually like trying to become human.

NC: Which is harder than you'd think. Something about the rawness and the honesty of it, which we worked towards in improvisation—at the end of the day, we are not empty, physical flesh. Saying this is me, a human being, with my experiences, is to look at our own species.

SE: And maybe to shift the lens of the inquiry.

NC: And to bring this into a performance space in which we also, simultaneously, have actors and characters, is to highlight the elemental.

SE: The elemental—like the elements?

NC: Yes. At the end of the day, we weren't trying to embody things so much as strip them down. Until we ourselves were elements.

SE: Is that what it was like to perform that opening monologue?

NC: I actually had two opening monologues, and the first was rooted, for me, in the second. Which was the first moment to acknowledge the audience. In my mind, that was my first stage direction. To be present on the stage and acknowledge the audience. And then to launch into embodiment. It was playful; it had many elements of what we decided to call "clowning," which was full of characterizations, over-the-top characterizations. I was attempting to latch onto non-psychological impulses that would carry me through that element or being—whether it be sniffing, or sexual desire, or desire to infect—that was probably the most anthropomorphized, the mosquito. That one was an action, the action of addressing the audience in a way that one crazed, aggressive human would—that one was less literal.

SE: What I'm hearing from both of you is that you were using multiple modes.

MA: Yes, but Carla and Lewis had very different roles. For [Nick], it was about finding the Nick in all of it, but for me it wasn't about finding the Meng in all of it, but finding the Carla. Carla and Lewis were very specific characters. For me, it was more about finding out how to perform actions and how to make the actions more exciting. To begin with, they're kind of volatile, and as characters, they embody lots of different things. But in a very different way than Nick did. In some way it's similar because it was in many different modes. It wasn't just physical, sometimes it was something that the audience might not even have seen—it was an internal embodiment that affected the performance in the moment. That happened in tiny moments in the play: we'd have a flash of embodiment. For instance, when we saw Daniel as the wall, we, for a moment, became a wall as well. That was the characters: the characters were very malleable, they were not really humans, but they resembled humans most of the time, and then they transformed briefly. But it wasn't about, "why are you doing this?" For them there was no "why." It was important not to have a "why." It was just what they were doing. It was very manic, actually. The thing is, the way they're written, they're very close to being just bratty teenagers. But the way I dealt with that was to think of the brattiness as a kind of becoming, like a becoming-teenager as another element, another form. The important thing was to make it not intentional. It's important that you can't follow their trains of thought, because they don't really have them. It's not psychological.

SE: What do you think were the successes of the production, in terms of the questions we laid out? Both the critical questions and the theatrical questions.

NC: The beginning. The beginning felt vast. The broadness and blurriness and grayness of the first quarter of the play. Being in six different places at once. There's something about that overture that introduced the world very successfully—introduced these different pieces that are all interconnected, that are right up against each other—the theatre, the landscape, the elements, the different species.

"Writing Ecocide"

Shonni Enelow, Playwright

I wrote *Carla and Lewis* in stages, in dialogue with the three workshops of the Ecocide Project. The first draft of the play came out of the unofficial first phase of the process, the series of conversations between myself, Una Chaudhuri, Fritz Ertl, and Josh Hoglund, and it focused on the political problem of representing ecocide as well as the perennial ethical, political, and representational problem articulated by Gayatri Spivak as "can the subaltern speak?"[3] In my play, provisionally called *McCloud* after Robert Altman's 1970 film *Brewster McCloud*, about a young man obsessed with birds who builds wings and (suicidally) flies around the Houston Astrodome, the subaltern in question was both the human subaltern Spivak addresses and the non-human subaltern that animal studies and ecocriticism have more recently brought to our attention. Can the non-human speak? *McCloud* was about two artists named Carla and Lewis who build a nightmarish art installation in the Houston Astrodome, of exotic animals made to perform as human psychiatric illnesses. From that first draft, only Carla and Lewis, the punk butterfly artists, remained, and with them, the meta-commentary on our own art-making. This decision to thematize our own process of making the play within the play came out of our desire to foreground the present-ness of the process and also to put ourselves on the line (other research theatre projects have taken the same tact: both of Steven Drukman's plays have meta-theatrical elements).

Looking back on it, I can see that my first impulse was to write a tight, weird, explicitly anti-utopian play: a play about cruelty, in both the Artaudian sense and the common sense, and abjection. It was always clear to me that this would have to be a pretty biting play if we were going to avoid earthiness. It had to be the opposite of earthy, which to me meant irony, sharpness, even archness: anti-Nature. But in the first workshop at NYU, the most striking pieces the actors brought into the room suggested something else, something less human than irony, and more spatially and conceptually expansive than the story I had imagined: a physical and psychic deterritorialization that we began to call "becoming-landscape." Deleuzean language is the best way I know to describe these unhingings of body and humanness (i.e., interiority, coherent psychology, selfhood), these exteriorized acts in which the fullness and density of the room was a lively, and uncanny, presence, intermeshed with the bodies of the performers in startling, layered ways. A mouth became a door; five bodies became a forest; an iguana-dancer turned our studio into an empty supermarket.

The question for me was how to write in a way that would catalyze these becomings-landscape but not determine them. What kind of language could do that? This was my question during the fall workshop at The Invisible Dog in Brooklyn, an art center in a former factory building with its own particularly dense kind of vastness. Before writing anything myself, I collected passages from some of our most piquant literary source texts—especially Aira's *Episode in the Life of a Landscape Painter* and Marian Engel's *Bear*—as well as more mundane anthropological and journalistic sources to give to actors as prompts, to see what was sparked by different types of language. I found that concrete, non-narrative language was the most productive. For instance, this passage from Aira:

> It was like wandering from room to room at a party, from the living room to the dining room, from the bedroom to the library, from the laundry to the balcony, all full of noisy, happy, more or less drunk guests, looking for a place to cuddle or trying to find the host to ask him for more beer. Except that it was a house without doors or windows or wall, made of air and distance and echoes, of colors and landforms.[4]

If the language was too abstract, it got fuzzy and dull. If it was too narrative, it trapped the performers in a story-telling mode. If it was too direct and literal, there was no room for discovery. If it was too oblique, I wasn't helping them at all.

But how to balance the abstract and the concrete, the poetic and the literal, the metaphors and the materials? Also (and this doesn't map easily onto the previous pairings), the human and the non-human: the non-human materialism of the theatre that we wanted to emphasize and the unavoidable charisma of the human body. Instead of tackling that problem directly, it was helpful for me to turn to a more grounded series of questions: where should the audience look? What should they pay attention to? How could I shift the lens and keep shifting it?

The heat island composition, reprinted in Fritz Ertl's chapter of this book, came out of these questions and our becoming-landscape experiments. Writing it, I was inspired by the detritus-strewn landscapes of post-modern drama, specifically those of Heiner Müller, from whose play *Hamletmachine* I took the Ophelia imagery and the line "SNOW ON HER LIPS," reading the famous final image of Müller's play as a climate change nightmare.[5] In Müller's scene, set in the deep sea, where "fish, debris, dead bodies and limbs float by," Ophelia speaks directly to the audience, describing in present tense her destruction of the world ("I choke between my thighs the world I gave birth to"), while physically, on stage, men in white smocks wrap her in gauze. Müller draws our attention to the limits of performative speech acts on the stage (challenging, as the entire play does, Brecht's theories of active spectatorship), as well as the zombie-like power of language to resonate beyond its (dubious) material efficacy; the audience is complicit in her smothering, but also, simultaneously, interpolated into her battle cry ("Down with the happiness of submission").[6] In that section, I rewrote the victim/perpetrator dynamic of Müller's scene—everyone on stage is Ophelia, and everyone on stage is watching her—but more importantly, I explored what in Müller's text remains latent: the unacknowledged non-human actant (Bruno Latour's term) of the water. Foregrounding this other model of action further unseats the binary between active and passive, actor and spectator, subject and object: after all, Ophelia doesn't drown Ophelia, the water does. More precisely, Ophelia enters into an assemblage with the water, acting in collaboration with it. How might our theatrical tactics change if we thought of actors and spectators as a similar kind of assemblage?

This shift in the terms of the active/actant/actor, inert/object/spectator binary was one way I tried to write against the hermetic tendencies of modernism. When we in the Ecocide Project said "becoming-landscape," we meant something somewhat different by landscape than did Gertrude Stein, who valued landscape as a theatrical model because a landscape "does not have to make acquaintance."[7] Our landscape *did* make acquaintance: the actors playing Landscape speak directly to the audience, alternately coy and belligerent. However, in revising Stein's landscape theatre, we also stayed true to her dislike of presumed familiarity and emotional arm-twisting, her insistence on maintaining the separateness of the stage. All four of us main collaborators have always said that we hate "interactive" theatre; I like that we chose to tackle that revulsion head on. Instead of "breaking down" the boundaries between actor and audience, which none of us thinks is very productive, we took it as a given that there are differences (not only unavoidable but desirable) between the actors and the audience, but also at least two things shared: we are all in the same space, and we are all of the same species. In the play, I tried to get at this latter, uncannier, easier-to-ignore correspondence (species), by critiquing the "interactive" model, which I did by satirizing the veiled aggression inside its coercive identifications: "I want to eat everything you eat."

I also meant this audacious making-acquaintance to evoke the hypocritical, and phobic, response of the bourgeois American subject to the rise of Asian capitalism: don't do as we do, don't consume like we consume. That Americans both cite Asia as today's prime example of climate change's destruction ("ground zero," as Elsa puts it in the play) *and* present it as the reason global action to combat climate change is impossible (because the Chinese won't stop buying cars) is a poignantly Orientalist irony: Asians are both powerless victims, to be protected, and uncomprehending perpetrators, to be enlightened. This neo-colonial subtext finds expression in the play through Kamna, Amina's double, who rejects Elsa's ideas about her life and agency and her efforts to pin climate change on Bangladesh in either fashion: "if there is a face of climate change," she tells Elsa, it is not Amina, the fantasized subaltern victim, but "the fender of the American SUV."

I didn't set out to make all of the characters in the play female, but I embraced it once I realized I had done so. A feminist impulse, certainly, and also a way of foreclosing heterosexual readings of intimacy. Also, I was interested in doubles, and the doublings and even becomings inside female relationships are something I keep coming back to as a playwright. Carla and Lewis's patois was inspired by Beckett but also by the uncanny intimacy and intersubjective pliability of transgressive little girls. Just as Amina and Kamna are doubles, too, so are Elsa and Bronwyn, the Scientist: both well-intentioned, earnest, enlightened investigators, they both try to arrange, categorize, and contain their research subjects, and are deeply frustrated when those subjects evade capture. All six characters (Carla and Lewis, Amina and Kamna, Elsa and Bronwyn) mirror and mesh with each other, co-evolving in response to the landscape and its stimuli.

In composing the play, I tried to incorporate the following components:

1. The landscape, including both the here and now and the there and elsewhere of theatre (articulated in the stage directions and activated by the landscape/actors)
2. The conceptual challenge of representing climate change (worked through in the story itself)
3. The aesthetic and ethic challenge of "can the (non-human and human) subaltern(s) speak?" (worked through in the doubled roles of Amina and Kamna)
4. Evolutionary character (represented by Carla and Lewis, punk butterflies)

One of the questions that remained was how much the elements on this list should fit together: how much of a coherent whole should this play be? Once we began the third workshop, immediately preceding the production, my job became less exploratory, and more about making things work, giving the production team a script they could understand and use. The script became more narrative and less open. This was a frustrating transition for me as a writer: I felt a great responsibility to the group to make something workable that also did justice to the depth and breadth of our research experiments. To be honest, I never felt wholly satisfied with the way that the play narrativized our critical questions; I'm not sure narrative can be "just" another element in a work like this. It might have been our pedagogic impulse that led us to greater narrative structure and simplicity: we wanted to make ourselves clear, and I'm not sure that that's what a work of art should do.

But this is one of the tensions built into the Research Theatre process, between the desire to make art and the desire to gain knowledge. We were trying, in other words, to be both Carla and Lewis and Elsa and Bronwyn, but in the end, we were more like Elsa and Bronwyn, trying to control our unruly research subject and fix in inside a structure we could understand. Except, of course, that just like Elsa, we failed: fruitfully and rather beautifully, if I may say so. This is the power of Elsa's final monologue. She loses control; we did, too. What this reveals in terms of the process, I think, is that in order for this tension inside research theatre to remain productive, it's important to resist thinking of the play and the production as the "result" of the research: it's another stage of it. What this meant for me as a writer was that I had to let go some of my own aesthetic interests, and certainly my sense of artistic ownership over the results. Jean Genet once told an interviewer that "the closer a work of art is to perfection, the more it is enclosed within itself".⁸ This play is not enclosed within itself, and whether or not it can stand on its own is ultimately, for me, beside the point. In this project, the research exceeded what we could put into a short play; as Nick resonantly put it in our interview, this play is just the tip of the iceberg.

The irony of that image, of course, is that the tip of the iceberg is literally melting more and more each day: what will we researchers and artists do, I wonder, when we no longer have icebergs as metaphors for our work's inadequacies? On the one hand, there is tremendous urgency to this problem: we have to figure something out quickly, or the icebergs will melt, with zero apologies to our artistic and philosophical limitations. On the other hand, if this is the problem that defines our era, the Anthropocene, it may well be what we spend our lives thinking about. And perhaps the very fact that we use the image of the iceberg to convey the limits of our understanding indicates why this problem is far larger than one of government policy and capitalist consumption habits. I hope this play *is* the tip of the iceberg, and that other artists and thinkers will build on what we've learned and bring into view more of it.

Notes

1   Sally McKay, "The Affect of Animated GIFs (Tom Moody, Petra Cortright, Lorna Mills)," *Arts & Education*, accessed November 22, 2013, artsandeducation.net/paper/the-affect-of-animated-gifs-tom-moody-petra-cortright-lorna-mills.
2   Interviewed by Shonni Enelow, September 5, 2013.
3   See Gayatri Chakravorty Spivak, *A Critique of Postcolonial Reason: Towards a History of the Vanishing Present* (Cambridge, MA: Harvard University Press, 1999).
4   César Aira, *An Episode in the Life of a Landscape Painter*, trans. Chris Andrews (New York: New Directions, 2006), 66.
5   Heiner Müller, *Hamletmachine and Other Texts for the Stage*, ed. and trans. Carl Weber (New York: Performing Arts Journal Publications, 1984), 54.
6   Ibid., 58.
7   Gertrude Stein, *Last Operas and Plays*, ed. Carl Van Vechten (Baltimore, MD: Johns Hopkins University Press, 1995), XLVI.
8   Jean Genet, *The Declared Enemy: Texts and Interviews*, ed. Albert Dichy, trans. Jeff Fort (Stanford, CA: Stanford University Press, 2004), 48.

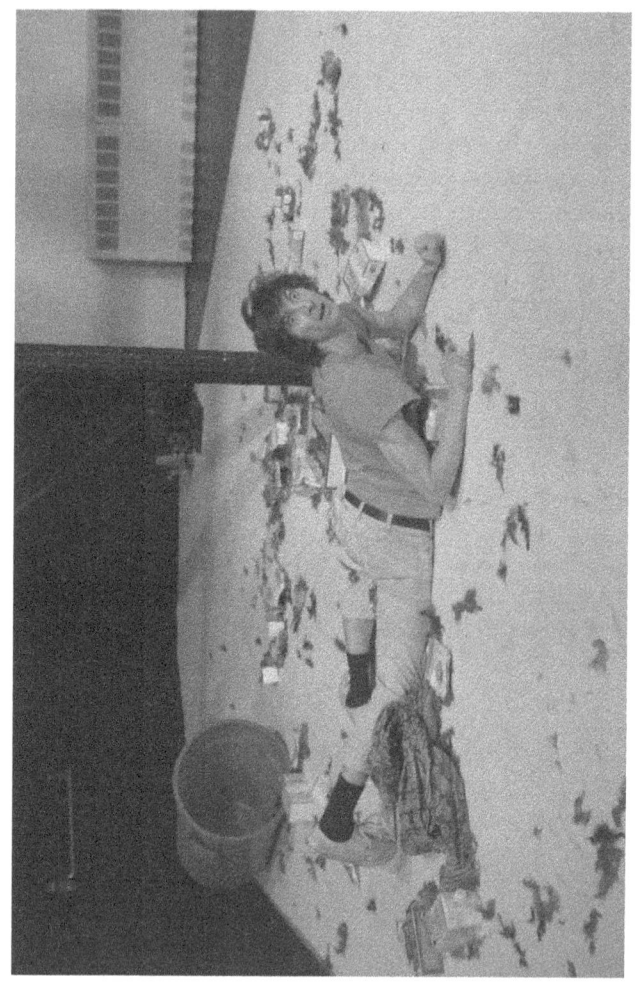

FIGURE 4.2 "Out of the mud come: Crocodiles. Malaria. Rotting wood." Nick Cregor as Landscape (#4745). Photo by Louisa Marie Summer

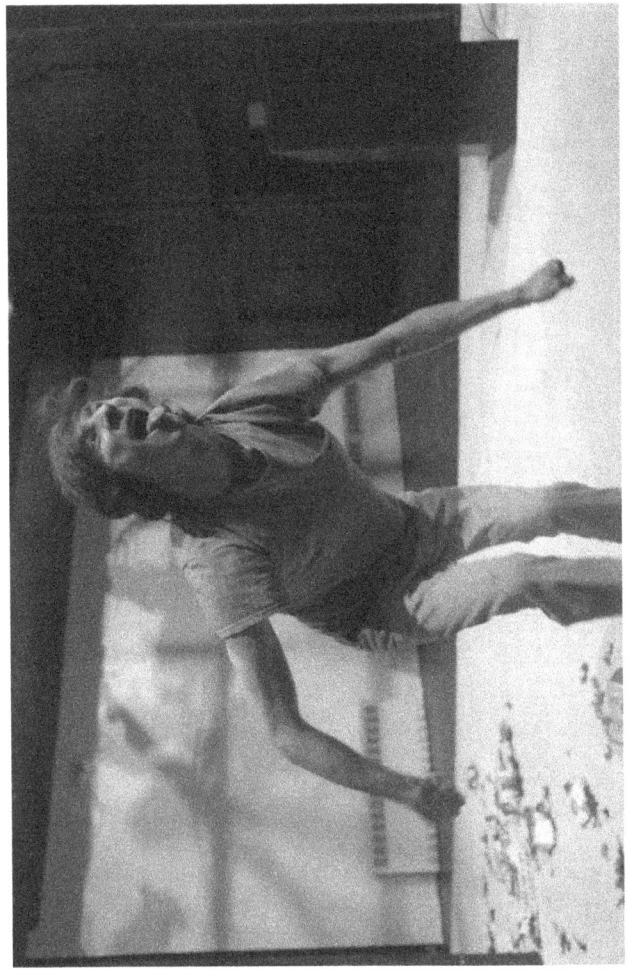

FIGURE 4.3 "What time did you eat dinner?" Nick Cregor "Makes Aquaintance" (#4761). Photo by Louisa Marie Summer

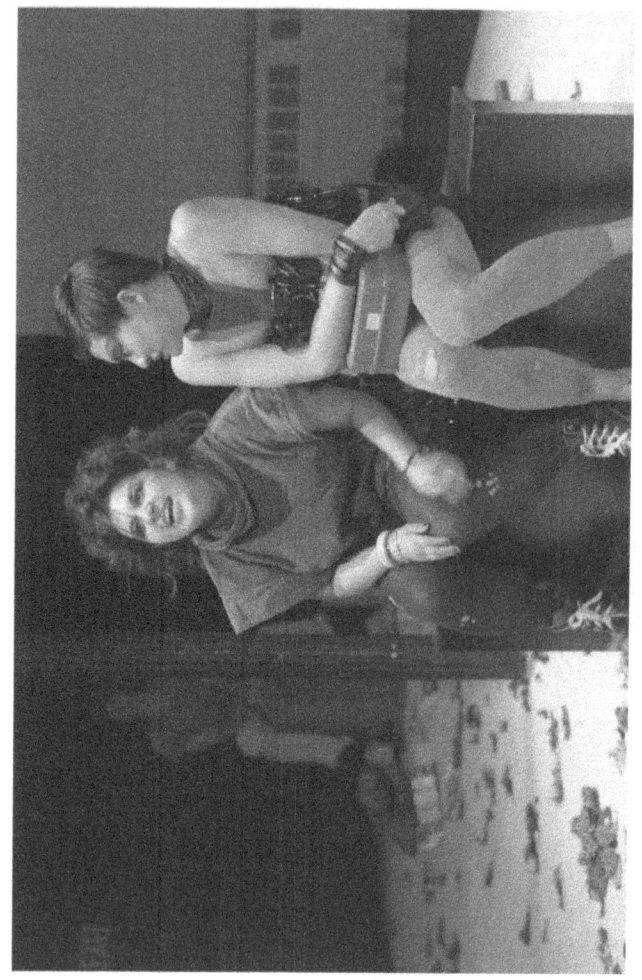

FIGURE 4.4 "So long we could feel our cells dividing." Kim Rosen and Meng Ai as Lewis and Carla (#4893). Photo by Louisa Marie Summer

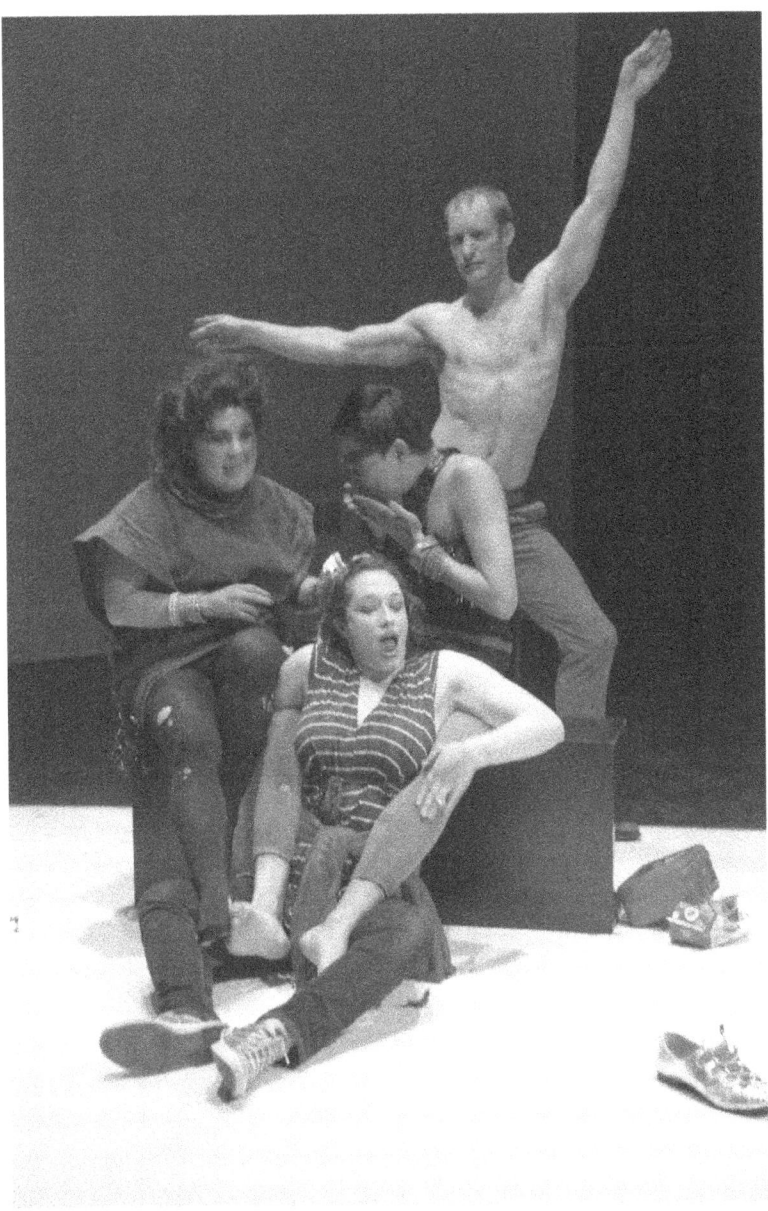

FIGURE 4.5 *Daniel Squire as Emotional Wall and Nicole Gardner as Mud (#4976). Photo by Louisa Marie Summer*

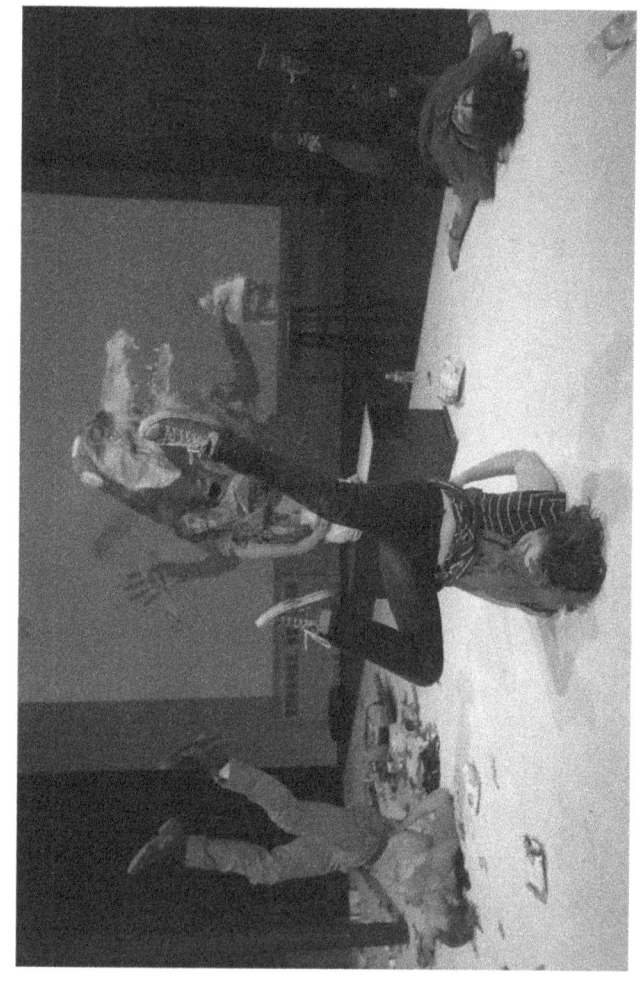

FIGURE 4.6 "I'll sleep in mud and die in mud, be born in mud." Nicole Gardner, Daniel Squire, Kim Rosen, and Animated Drawings (#5116). Photo by Louisa Marie Summer

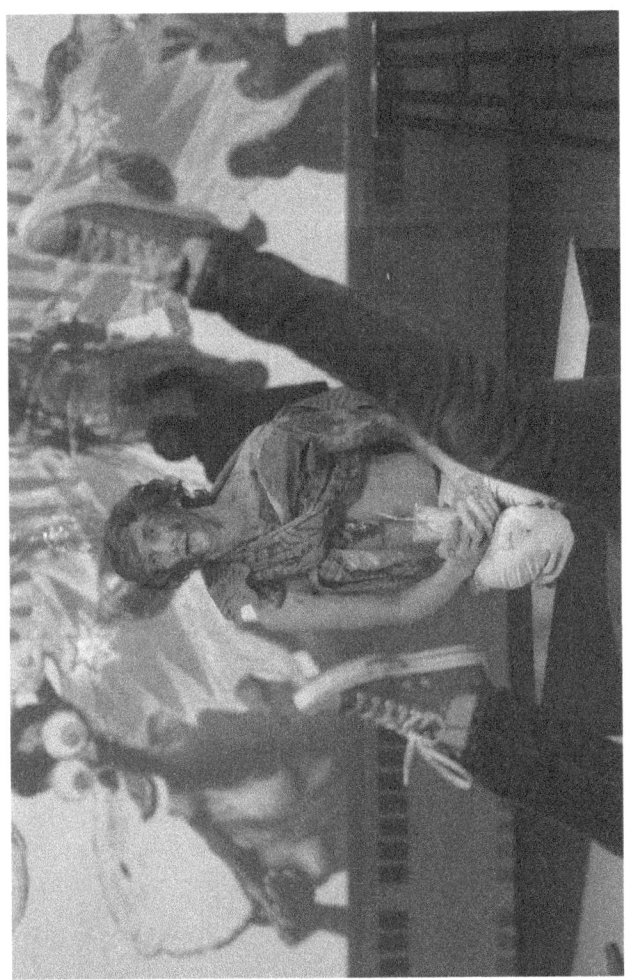

FIGURE 4.7 *Becoming-Landscape: Nick Cregor, cast members, and animated drawings* (#5115). Photo by Louisa Marie Summer

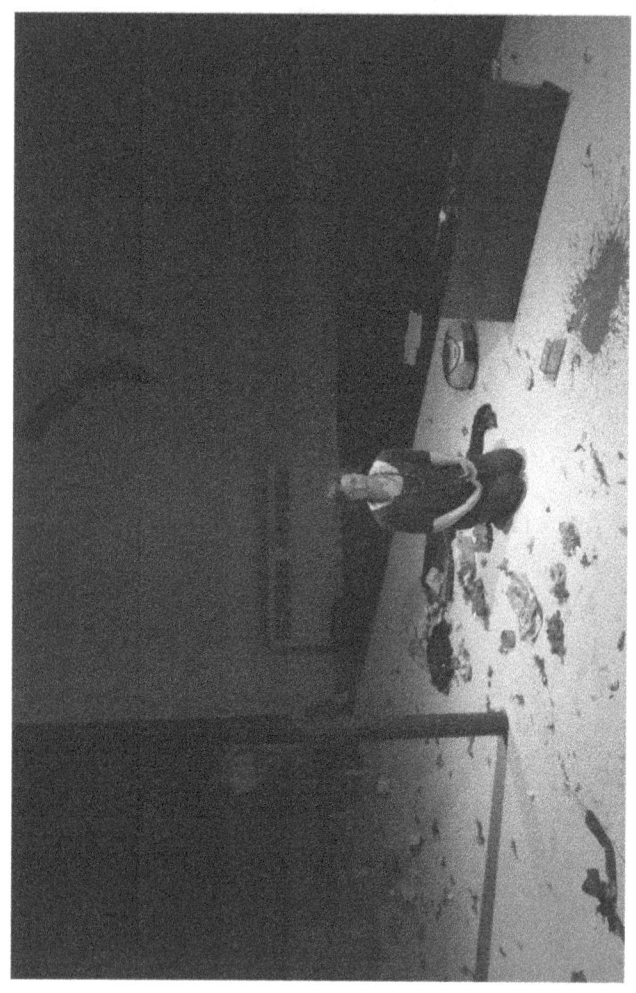

FIGURE 4.8 "And I say the crocodile is taking my sister, and I cry the salt of the marsh." Libby King as Elsa-Becoming-Amina (#5249). Photo by Louisa Marie Summer

5
Carla and Lewis

Shonni Enelow

Abstract: This chapter includes the full text of the play *Carla and Lewis*, by Shonni Enelow. Two punk butterflies named Carla and Lewis land in the East Village apartment of curator Elsa Turner, who has conceived an interactive art installation that will speak to the human face of climate change. What she doesn't realize is that the mud of Bangladesh, which she imagines to be so far away, is in fact soaking her apartment walls and creating its own acrimonious ecosystem right under her nose. Carla and Lewis, who don't play by the rules of humanness, eventually force her to see the mud for what it is, transforming her world, and the theatre, with their audacious, disobedient, co-evolutionary mutations.

Keywords: art installation, Bangladesh, climate change, punk butterflies, refugee

Chaudhuri, Una and Enelow, Shonni. *Research Theatre, Climate Change, and the Ecocide Project: A Casebook.* New York: Palgrave Macmillan, 2014.
DOI: 10.1057/9781137396624.0009.

A theatre.
An actor is onstage.
Pours mud all over the floor.
Out of the mud come:
Crocodiles.
Malaria.
Rotting wood.
Rats, preening like birds.
Dead fish.
Computer parts.
Amina.

ELSA: Hello, my name is Elsa Turner and I'm an independent curator of the fine arts. I'm very pleased to speak to you tonight about my new installation project, tentatively titled "The Amina Project," which will premiere this summer in New York. I began developing the Amina Project in the spring of 2010, when I came across an article in *The New York Times* about the increasingly dire situation of climate change refugees worldwide, and especially in Bangladesh, "ground zero" for the disaster. Rising ocean levels will wipe out more cultivated land in Bangladesh than anywhere else in the world. In the next twenty years, as many as 15 million people could be displaced.

In the article, there was a picture of a woman named Amina, who lives in the village of Gabura in southwest Bangladesh, and who received a fractured collarbone when a tidal wave came through the wall of her shack. When the journalist asked Amina why she and her husband didn't move, she replied, "we're all poor people. We don't have anywhere to go." And then her story breaks off. But I kept looking at that picture of Amina. I decided that somehow, Americans must hear her voice. And clearly, *The New York Times* is not enough. That was when I started developing my project.

The Amina Project will be a large-scale performance installation in which New York gallery-goers will have video conversations over Skype with climate refugees, including Amina herself, who will be present in a gallery in Bangladesh. Visual artists will be stationed in both galleries to respond—in drawings or paintings—to the video conversations taking place. When the performance is over, both the artists' visual responses and the conversations will be collected and published as a book.

I am happy to report that I have already found both a Bangladeshi artist and two American artists to take the project on. Finding the Bangladeshi was simple. After a few inquiries I found Kamna Banerjee, a feminist mixed media artist, who lives in Dhaka—that's the capital city of Bangladesh. Finding the New York artists was a bit more complicated.

The landscape shifts.
Becomes a city.
The elements of the city include:
A subway car that is also a Laundromat.
Rats, preening like birds.
Milk and potato chips.
Perpendicular movement: standing vertically, then folding like a screen.
Trees/illegal immigrants.

I only moved to the city recently and I'm not quite settled yet. Can I tell you—I had a very unpleasant experience last week. I saw a man keel over in the subway. I tried to help him and he started screaming at me in another language and the next thing I knew, my briefcase and laptop were gone. Does that sort of thing happen all the time here or was I just unlucky? On Tuesday all my underwear was stolen from the dryer at the Laundromat. What? (*as if responding to a question from the audience*). Oh, from New Haven. Yes, it's very different.

Anyways, after weeks of interviewing any New York artist I could get my hands on, I finally got wind of something interesting. Apparently there were two American artists based in Berlin who'd gotten a lot of attention for a durational performance piece they did in their apartment. For a year, they stayed indoors, drawing naïve colored pencil pictures on their walls, pictures taken from whatever happened to be on television. They left their front door open and anyone could come watch them. Eventually, there were more than fifty people at a time. It was such a brilliant interpretation of the society of the spectacle—a meditation on gender, the image, the body in time. Their names were Carla and Lewis.

Carla and Lewis appear.

ACTOR 1: Did you take a train to this theatre? Did you see any rats? Did you see any star-nosed moles? Did you walk to this theatre? How was your epifascial system? How was gravity? Were you able to remove enough oxygen from the air? Did you feel the vapor saturation of the air? Did your breathing make you lose too much water? Do you eat enough vitamins to provide for healthy hair and nails? Do you make faces of pleasure or displeasure as you walk? Did the passive diffusion of gases to your heart provide enough for total circulation?

ACTOR 2: What time did you eat dinner? (These lines should be improvised) I ate (whatever Actor ate).

CARLA: You want to talk about eating?

LEWIS: Speak up! Look up!

CARLA: I haven't eaten anything since Berlin.

LEWIS: We made red lentils at 5 am and ate them twice a day until they ran out.

CARLA: With pepper and beer.

LEWIS: And the old woman who brought us buckets of vegetables.

CARLA: We didn't cook the vegetables.

LEWIS: We wore them on our belt loops.

CARLA: She was a farmer.

LEWIS: You think I'm kidding?

CARLA: Lewis isn't kidding.

LEWIS: I'm Lewis.

CARLA: I'm Carla.

LEWIS: We were living in Berlin.

CARLA: We had an enormous apartment and a sink like a bathtub and no refrigerator and a mattress we found in the street.

LEWIS: It was great.

CARLA: Why'd we leave?

Lighting shift.
One of the Actors puts on a headscarf to play Kamna.

KAMNA: Dear Ms. Turner, thank you for your email. I would be delighted to collaborate with you on an installation project, and would be very glad to hear more about what you have in mind for the gallery here in Dhaka. I must first admit, right off, that I don't know anyone like Amina. I live in the city. I attended the university. I have never been to the villages of the south. But I will do my best to make the appropriate inquiries to see if Amina can be found.

ELSA: Dear Ms. Banerjee, or Kamna, if I may: I am so pleased that you are interested in the project and I look forward to discussing it with you further. The first thing to do is to find Amina. I imagine it wouldn't be too much trouble to locate her and her shack. Or if not her, specifically, another woman in an analogous situation. Let me stress that this first step is of ultimate importance: I want to find actual refugees. I'm attaching the *New York Times* article, with Amina's picture. Perhaps this will be of use. Please keep me abreast of your progress while I deal with things on my end here in New York.

The mud landscape and the city landscape collide.

ACTOR 1: HUMANS! LEAVE YOUR HOMES!

CARLA: WE WILL WE DID!

LEWIS: WE MOVED BACK TO NEW YORK!

The weather changes.

ACTOR 1: Hello people, hello kind, wise, rational, self-aware people, welcome to this play. This is the scene in which we introduce the world and its landscape, which is of course this theatre: its flora and fauna, its technological machinations, and its natural and man-made ornamentation. Over here, you see, there is a stairway, otherwise known as risers, on which there are approximately 80 metal chairs. The metal is aluminum, the silvery white member of the boron group of chemical elements. It is the most common element in the earth's crust. Plants ingest it in their food, the soil, as do animals, who ingest plants. You ingest it.

ACTOR 2: You put it under your arms if you use antiperspirant. I don't. Do you think I smell?

ACTOR 1: Now, these four bodies you should pay attention to. This is Elsa. Humans like her like to understand and influence their environment. She is also the only living species of bipedal primates in the great ape family. So congratulations to Elsa. This actor over here will be playing two roles: Kamna, a Bangladeshi artist, and later in the play, a climate change scientist with a bad haircut. These two are punk butterflies. They are the most important.

ACTOR 2: Carla and Lewis.

ACTOR 1: The American artists Elsa heard about. They've been migrating from Berlin to New York, as they do every couple years when they run out of food, and they are exhausted.

The subway platform that is also a Laundromat.

CARLA: The plane was revolting. Lewis threw up twice in the bathroom.

LEWIS: Out of sheer disgust!

CARLA: Finally we landed.

LEWIS: We smelled like peanuts.

CARLA: And novacane.

LEWIS: And mayonnaise.

CARLA: And sweat.

LEWIS: We didn't have any baggage.

CARLA: Finally we got to the subway station.

LEWIS: It was dusk.

A sound.

CARLA: I hear something.

LEWIS: What is that?

CARLA: Is it the radiator?

LEWIS: Radiator?

CARLA: Are those crickets?

LEWIS: Crickets?

CARLA: In New York?

LEWIS: It's a horrible noise.

CARLA: I think it's wonderful.

LEWIS: Horrible.

CARLA: Wonderful.

LEWIS: Horrible.

CARLA: Wonderful.

LEWIS: DESPICABLE!

CARLA: PHENOMENAL!

LEWIS: Sometimes we do this, we contradict each other, it's not a big deal.

CARLA: WONDERFUL!

LEWIS: That's enough.

CARLA: No it isn't.

They smile.

CARLA: Sometimes the trees have emotions.

LEWIS: That's true, actually.

CARLA: They have more emotions than we do.

LEWIS: We actually have very few emotions.

CARLA: Anger, satisfaction, desire.

LEWIS: Disgust.

CARLA: That's about it.

LEWIS: We've narrowed them down.

CARLA: We like to simplify.

LEWIS: We narrowed down everything when we were in Germany.

CARLA: We had a couple of spoons and a mattress and that was about it.

LEWIS: We knew very few people.

CARLA: We kept to ourselves.

LEWIS: And played our records.

CARLA: And drew little pictures on the walls of our apartment, little stick-figure pictures of us doing things.

LEWIS: There was a drip in the bathroom.
CARLA: It became like the crickets.
LEWIS: Or like people talking.
CARLA: Like people talking. Like crickets talking.
LEWIS: Very expressive drip. Many emotions.
CARLA: More than us.
LEWIS: More than we had.
CARLA: Many more.
LEWIS: And I developed a weird tic where I would think I would see something out of the corner of my eye.
CARLA: She would jerk her head around, like that, all the time.
LEWIS: But there wasn't anything.
CARLA: Because all we had were spoons.
LEWIS: Some duct tape.
CARLA: And colored pencils, for drawing on the walls.
LEWIS: We went out sometimes. At least once a week, usually.
CARLA: We would walk to the park extremely slowly.
LEWIS: Or sometimes we'd run.
CARLA: We'd run.
LEWIS: We'd run to the park and around the park and to the canal and into the water into the water of the canal!
CARLA: Like there was someone chasing us!
LEWIS: Well there was someone chasing us!
CARLA: Could have been anyone!
LEWIS: A man maybe! A sweet old man with a cane!
CARLA: EW DISGUSTING!
LEWIS: I wanted to get him!
CARLA: I wanted to eat him!
LEWIS: We'd run until we collapsed!

They collapse.

    CARLA: We waited on the outdoor platform.
    LEWIS: By JFK, that outdoor platform.
    CARLA: For a ridiculously long time.
    LEWIS: So long we could watch our nails grow.

CARLA: So long we could see spots in the sky were the sun had moved.

LEWIS: So long we could feel our cells dividing.

CARLA: Where is this fucking train?

Pause.

LEWIS: Then we got the call.

CARLA: Ring, ring.

LEWIS: Hello? (*Note: she is not using a phone of any kind*)

ELSA: Hello, may I ask who is speaking?

LEWIS: Hello, may I ask who is speaking?

CARLA: Who is it?

ELSA: Yes, of course. I'm sorry. My name is Elsa Turner and I'm a curator.

LEWIS: A curator.

CARLA: Really?

LEWIS: Really?

ELSA: Yes, I got your phone number from Siri at the cropland website.

LEWIS: How did you get this number?

ELSA: I got this number from Siri Copberg at the cropland website.

LEWIS: What website?

ELSA: Siri, Siri Copberg. The cropland—no, I'm sorry, the rangeland website.

LEWIS: Oh Siri Cropberg at the Rangeland.

CARLA: (*to audience*) Siri Cropberg at the Berlin Rangeland.

LEWIS: Who are you?

CARLA: She's a curator.

ELSA: I'm recently independent. I mean I'm a curator. I used to work at—

CARLA: Lots of places that we know.

ELSA: Yes anyways I've been working independently for the last year or so and I actually have an opportunity I thought perhaps—

CARLA: It was an opportunity that she thought perhaps—

ELSA:—would be attractive to you.

LEWIS: Oh absolutely, we'll take whatever we can get!

CARLA: Ha ha!

LEWIS: I mean for eating, we need to eat you know.

ELSA: So you'd like to eat for dinner?

LEWIS: Sure we like to eat for dinner!

ELSA: I meant meet for dinner.

CARLA: Oh absolutely.

ELSA: Great, that's—

LEWIS: Wonderful we're on a subway platform at this moment.

ELSA: Oh you are?

LEWIS: We're hungry.

CARLA: We'll be there in 40.

ELSA: Where?

LEWIS: Wherever you want!

Lighting shift.

CARLA: We took the commission.

LEWIS: We needed the money.

CARLA: We had no choice.

LEWIS: We had to go somewhere.

CARLA: She even told us we could stay in her apartment for a few days.

LEWIS: What does this lady want with us?

CARLA: What do you think?

They smile.

LEWIS: I like older women.

CARLA: Their skin is mushier.

LEWIS: Separates off the bone easier.

CARLA: I like curators.

LEWIS: I like places to stay.

CARLA: I like potato chips, blue light, and submarines.

LEWIS: I like whole milk, Visine eye drops, and biting.

Lighting shift.

KAMNA: Dear Elsa, I am writing to keep you informed of my progress. I've spent quite a bit of time researching the science of climate change. What I've found so far fascinates me. What's struck me is how de-centered the phenomenon is. There is no origin, no unified symbol, no one meaning to the whole thing. And it's as if this very structure has closed it off from representation. I have no idea how to make art about climate change. Sure, we could find an image—like you did—of a person—like you did—"Amina." But to me that doesn't seem right. Climate change is enormous, it's tiny, it's impossible, it's happening—all the way up and all the way down. This has been my guiding idea in developing the visual idiom for the gallery drawings. I have not yet found the time to go to the south and find Amina. I am getting married next month and in the process of moving apartments. In fact I wonder if Amina is really worth the trouble. I doubt she will have anything interesting to say. She was poor before and she is still poor. She was wretched before and she is still wretched.

ELSA: Dear Kamna, how wonderful to hear about your marriage. I can only imagine how full of excitement—and, perhaps, trepidation?—you must be. And what a coincidence about your housing stress—I've been having tremendous apartment stress too. The building management refuses to repair a major leak, and I fear I may have to move. But moving is such a terrible hassle, isn't it? I would love to hear more about your experience. This is the kind of open communication I hope we can impart to the participants of our installation project.

KAMNA: Dear Ms. Turner, thank you for your good wishes on my impending marriage. It was arranged by my grandparents. But you have not responded to my ideas about the project. I am certain it has nothing to do with Amina.

Elsa's apartment.
The floor is covered in mud.

CARLA: What is this place?

ELSA: What do you mean?

LEWIS: The floor?

CARLA: I love it!

ELSA: Oh, I know. Over the summer there was construction and they ruined the insulation. I've called and called but the landlord won't fix it.

CARLA: You live like this?

ELSA: It's a little damp, I know. It's just the insulation. I'm working on getting it fixed.

CARLA: Do you like to feel your toes sinking into it?

ELSA: I usually do ask people to take off their shoes.

LEWIS: Then we agree.

CARLA: Do you have any chips?

ELSA: I think so, in the kitchen.

CARLA: Well where else would they be?

ELSA: Nowhere.

LEWIS: (*from kitchen*) Found some!

ELSA: Good. Make yourselves at home.

Over the next dialogue, Carla and Lewis make themselves at home.

CARLA: This wall reminds me of our apartment in Germany.

ELSA: That wall?

CARLA: Yes, this one.

LEWIS: Very expressive wall. Many emotions.

CARLA: Did we tell you about our apartment in Germany?

LEWIS: Our home.

ELSA: It's hard to move, isn't it. I hope you'll feel at home here.

CARLA: What do you want us to do with you?

ELSA: You mean, with the work?

LEWIS: If you say so.

ELSA: I think you should start by developing the visual vocabulary.

CARLA: Absolutely.

LEWIS: Visual vocabulary.

ELSA: Right. But it would help for you to know the facts, wouldn't it?

LEWIS: The facts.

ELSA: I've done lots of research. My neighbor is a scientist and she's been telling me everything. She's been working in a lab for the last twenty years studying the effects of changing water temperatures on a certain freshwater crustacean.

CARLA: Crustacean.

ELSA: Yes! Fascinating. Rising temperatures affect every level of the ecosystem. It's all connected.

LEWIS: So what happens?

ELSA: Excuse me?

CARLA: To the crustacean.

LEWIS: You said she was studying effects.

ELSA: It gets damaged.

CARLA: How?

ELSA: I don't know. I think it dies.

LEWIS: Or maybe mutates.

CARLA: Might just mutate.

LEWIS: Maybe it has to leave its home.

CARLA: Terrible.

LEWIS: Traumatic.

CARLA: We miss Germany.

ELSA: (*sympathetically*) Oh I see.

ACTOR 2: HUMANS! LEAVE YOUR HOMES!

The weather changes.

ELSA: I'll also put you in touch with the Bangladeshi artist, once I'm sure who that'll be. I'm in talks right now with an artist from Dhaka—that's the capital of Bangladesh—named Kamna Banerjee. But right now, I think it's best if you two work on your own. I want you to feel free to experiment with lots of visual idioms before we settle on a collaboration.

CARLA: You have another girl?

LEWIS: Where is she?

ELSA: Who—Kamna? She's a Bangladeshi artist. She's in Dhaka—that's the capital of Bangladesh. You can't see her. Besides, I'm not sure she's the right person for the job.

LEWIS: Why not?

ELSA: Well, she has a different take on the whole thing.

CARLA: What kind of different take?

ELSA: Let's just say it's different.

LEWIS: It's different.

ELSA: Right. But the key to the whole thing is Amina. Remember I told you about her at dinner?

CARLA: That other girl.

LEWIS: With the broken collarbone.

CARLA: So many girls.

ELSA: Let me show you a picture.

Rotting wood.
Dead fish.
Computer parts.
Amina.

CARLA: Look at her.

LEWIS: Look at her.

CARLA: Look at her!

LEWIS: Look at her!
CARLA: This mud!
LEWIS: Delicious!
CARLA: In between my toes!
LEWIS: Crumbly!
CARLA: Squirmy!
LEWIS: Like rotting!
CARLA: Like rotting wood!
LEWIS: Squirmy!
CARLA: Like sex!
LEWIS: Rotting mattress!
ELSA: That's not what she's thinking.
CARLA: My dress is soaked.
LEWIS: But I'm not moving.
CARLA: I'm not going anywhere.
LEWIS: I'm going to sit in this shithole.
CARLA: Until the end of time.
LEWIS: Until my head's underwater.
CARLA: And then I'll die.

Dead fish.

ELSA: What are you talking about?
LEWIS: I could live in water.
ELSA: I think it's important not to—
CARLA: Better than metal.
LEWIS: Dirt.
CARLA: Shack.
LEWIS: More space now.
CARLA: Water!
LEWIS: Toes!
ELSA: I think it's important not to assume that her life is anything like ours. I mean not assume until you actually talk to someone who has undergone what she's gone through.
CARLA: I cut her hair once while she was sleeping.

ELSA: What?

LEWIS: Remember when I did that?

CARLA: Yeah, I was so mad.

ELSA: Oh you mean—

LEWIS: It was kind of a good haircut though.

CARLA: What do you know about fashion.

LEWIS: I know tons about fashion.

CARLA: You know nothing about fashion. Look at what you're wearing.

ELSA: Listen, there's a line here I don't think we should cross.

CARLA: A line?

ELSA: Between making the story human and trivializing it.

LEWIS: Which one were we doing?

ELSA: I think you were trivializing it.

LEWIS: Oh good, I thought you were going to say the other.

ELSA: I just want to make sure that you understand the magnitude of this.

CARLA: The magnitude?

LEWIS: You mean, it's really big.

ELSA: It's an enormous issue. We're talking about peoples' lives.

LEWIS: Her life, you mean.

ELSA: Right, her. But she's representing lots. I mean look at all these people. We need to listen to them.

CARLA: How?

ELSA: Skype! Listen. I have an idea of how I want this all to go. Is it chilly in here or is it just me?

CARLA: The walls are wet.

ELSA: Oh, god, I know, that's just this—faulty insulation. The building management is totally negligent. I keep calling and calling and THEY JUST NEVER CALL ME BACK! I'm leaving messages two or three times a week and it's like they just tune me out! Just the crazy woman on the third floor! Listen, I say to them, there is WATER in the WALLS! I can FEEL IT! But the one time they show up, it was an exceptionally dry day and I couldn't—anyways. It's SO unbelievable! I really should file a complaint with the city.

LEWIS: You REALLY should!

CARLA: You should file a complaint!

LEWIS: You know they never listen to you unless you made a TON of noise.

CARLA: You want us to talk to them?

LEWIS: We can be really mean if we want to be.

CARLA: Just tell them you have two chicks up here recently returned from Germany!

LEWIS: Yeah, say this to them: ever BEEN to Germany?

ELSA: Maybe you're right.

CARLA: Have you lived here long?

ELSA: Just moved in a few months ago. Just moved here from New Haven.

CARLA: Why'd you move?

LEWIS: So awful to move.

ELSA: I was working at a gallery there, but I quit.

CARLA: Difference of opinion?

ELSA: Yes, how'd you know?

LEWIS: Oh we have lots of those.

ELSA: I didn't like the direction the gallery was going.

LEWIS: What direction do you like?

ELSA: Well—let's put it this way. I really think that if you aren't really making art out of reality—I mean the real situation at hand—you're just masturbating.

CARLA: You're not into that?

ELSA: Ha—no no, I just mean, I'm sick of talking about myself. I want to talk about people who are really suffering. Real people.

CARLA: So when do we get these people?

ELSA: What do you mean?

LEWIS: You said we get people to draw.

ELSA: Oh you mean—the Bangladeshis?

CARLA: Duh.

ELSA: That'll all be set up in the gallery.

CARLA: You don't have any here?

ELSA: No, but I have these pictures.

CARLA: Do you have any milk?

ELSA: I should in the kitchen.

CARLA: Well where else would it be?

ELSA: Nowhere.

Enter Scientist, played by Actor.
In the following scene, Carla and Lewis and Scientist circle each other.

SCIENTIST: Elsa?

ELSA: Bronwyn?

SCIENTIST: Oh, you have guests.

ELSA: These are the artists I told you about! Carla and Lewis. They just got in from Berlin.

SCIENTIST: Oh, hello. I'm the Scientist neighbor she told you about.

CARLA: Hello, Scientist neighbor.

SCIENTIST: You're Lewis?

LEWIS: That's Carla.

CARLA: I'm Carla.

ELSA: Bronwyn is the one studying freshwater crustaceans! I've been meaning to plan a meeting with the four of us! Bronwyn, can you stay?

SCIENTIST: Sure.

ELSA: We were just talking about the project.

CARLA: You kind of smell.

SCIENTIST: Excuse me?

LEWIS: Like chemicals.

CARLA: What's with your shoes.

ELSA: I was just telling Carla and Lewis about the concept. But you really are the expert on the—well the STAKES of the concept.

SCIENTIST: Lab shoes. Lab chemicals.

LEWIS: How revolting.

SCIENTIST: Well, excuse me. You're not exactly fresh daisies yourselves.

ELSA: I mean of course I understand the stakes myself, I just mean I'm still quite a neophyte when it comes to the science of the whole thing—

LEWIS: No, we're not daisies.

SCIENTIST: You kind of smell like dirt!

CARLA: Not surprising.

ELSA: I don't mean I don't know anything, I mean everyone knows something, the atmosphere and greenhouse gases and the evaporation and the—polar bears.

SCIENTIST: What?

ELSA: Nothing.

SCIENTIST: Do you know why I smell like chemicals? You know why I wear these shoes? You know what I do in that lab?

CARLA: Not particularly.

LEWIS: You could tell us.

SCIENTIST: I was just about to. I study climate change. The effects of climate change on the freshwater crustacean Daphnia. It's a water flea.

CARLA: So we've heard.

LEWIS: In fact we have a question or two about it.

SCIENTIST: Such as?

LEWIS: What happens to them exactly, Scientist? What kind of mutations are we talking about exactly?

SCIENTIST: What happens to the Daphnia? What happens? I'll tell you what happens. THE WORLD ENDS.

Pause.

ELSA: It's a very serious problem, just like I was saying, it's a problem of great magnitude—

SCIENTIST: The world ends. That's it. And there's nothing we can do about it except take a front row seat and watch it expire! You know I used to stand in the corner of Tompkins Square Park with a sign that says "The World is Ending," but after a while I realized people thought I was some kind of religious nut! I'm not a religious nut, I'm a scientist. Yep, I'm one of those dumpy scientists who work all day in labs.

ELSA: You're not dumpy.

SCIENTIST: Of course I am. I work all day in a lab with a bad haircut and orthopedic shoes. I love it. It's me. You need a calling in life. You two, have you found your calling?

LEWIS: Us? Sure.

SCIENTIST: What's it?

LEWIS: What?

SCIENTIST: What's it, your calling, what is it?

CARLA: Art.

SCIENTIST: Oh ART! How sweet.

LEWIS: You think?

SCIENTIST: Oh yeah. How cute and fun you little girls are. With those feathers and—what's that on your noses? That looks like duct tape. You're cute and sweet little girls with duct tape.

LEWIS: We're not little girls.

CARLA: You don't know us at all.

SCIENTIST: Oh don't get upset, baby, don't get upset! It's very hard, I'm sure, for little girls like you. You probably make very little money and meanwhile you've got to keep up those appearances, right? You've got to buy those scarves and feathers and duct tape and—what else—nail-polish and nail-polish remover and facial scrub and facial masking scrubs and facial scrubbing masks and lip gloss in multiple colors and oh I bet it adds up, doesn't it? I bet it all adds up and you don't even have husbands or husband-material lying around. It's hard, babies, it's very hard. You're talking to someone who knows. You think it's easy being a scientist with a bad haircut and orthopedic shoes? But you know why I do it? I do it for the children. I do it so the children can lift up their little faces to the bright light of day without fear of acid rain melting off those little faces. Actually what I do has nothing to do with acid rain. I just like the sound of acid rain. ACID RAIN. So dramatic. So '70s! You never hear anybody talking about acid rain anymore. Which makes you wonder. DID WE FORGET? Maybe we did, but we're not going to forget climate change, oh no! Let me tell you something about climate change, gorgeous, it starts all the way up top and it goes all the way down. Down to the freshwater crustacean *Daphnia pulicaria* and its single compound eye and its four to six appendages and its flat, transparent body. I study the temperature dependence of UV-induced DNA damage and repair in Daphnia, ladies. You know anything about THAT?

LEWIS: No.

SCIENTIST: I DIDN'T THINK SO! Nobody does but ME! I alone think about *Daphnia pulicaria*! ALL THE FUCKING TIME!

CARLA: Uh-oh.

SCIENTIST: Uh-oh, you say, she's getting shrill. I'm not getting shrill. Listen, I'll explain it to you very calmly and carefully. A daphnia is a water flea with a single, compound eye; four to six appendages attached to its thorax; and a flat, transparent body. It lives at the bottom of rivers. For most of the year, the population is entirely female and they reproduce asexually. And here I am with my lab coat and my nicotine habit sitting on my stool in that lab up on 14th street and torturing the *Daphnia pulicaria* eight to ten hours a day!

ELSA: I didn't know you smoked.

LEWIS: Well if it's so awful, Scientist, then why the hell do you do it!

SCIENTIST: WHY DO I DO IT? Well if you had let me FINISH you would have heard me say that the enzyme systems that repair UV-induced DNA damage are temperature dependent! So by exposing both live and dead Daphnia as well as raw DNA to UV-B, we are able to estimate the temperature dependence of *Daphnia pulicaria*'s DNA! And how quickly they will DIE when the water temperature RISES! And what that will DO to the rest of the WORLD! And the answer is: FINITO! OVER! DEAD! ENDED! You understand what I mean? You understand what I'm talking about?

The weather changes.

    LEWIS: Do we UNDERSTAND what you mean?

    CARLA: Do we KNOW what you're talking about?

    LEWIS: We aren't scared of crazy people!

    CARLA: We've just returned from Germany!

    LEWIS: Ever BEEN to Germany?

    CARLA: I've never seen more people with the look in their eyes of PURE CRAZY than in Germany!

    LEWIS: You don't scare me with your science, CRAZY CRAZY! Carla, does she scare you?

    CARLA: No she does NOT!

    LEWIS: We aren't scared of acid rain.

    CARLA: We ARE goddamn acid rain!

    LEWIS: It's coming, we're fucked, and we know it, we know it so well we can taste it. We're not scared of the polar ice caps melting!

    CARLA: We ARE the fucking polar ice caps melting!

    LEWIS: We are the hurricanes and the tsunamis and the flash floods and the fires. And we are the dead animals. The dead animals falling dead from the dead trees to the dead forest floor covered with other dead animals! So don't screech to us about science, GRANDMA! Take your fucking medication and leave us alone!

The weather changes.

    ELSA: Well this was very engaging.

    LEWIS: We'll go to our room now.

    CARLA: Pooped.

    LEWIS: Just plain pooped.

Carla and Lewis exit.

    SCIENTIST: They're idiots.

    ELSA: They're a little eccentric.

    SCIENTIST: You're letting them stay with you?

    ELSA: Just for a little while.

SCIENTIST: You're too nice.

ELSA: No, I'm not. It's my project they're working on.

SCIENTIST: That's what YOU think.

ELSA: I'm in touch with a Bangladeshi artist, did I tell you that? Kamna Banerjee—she's from Dhaka. That's the capital city of Bangladesh. I think it's very important that this project be conducted ethically. I mean I want the artists to feel they are full collaborators. It's important they feel they have total freedom.

SCIENTIST: I should go. Any luck with the management? It smells like mold in my bedroom.

ELSA: Are you kidding? They don't even call me back!

SCIENTIST: Me neither. It's ridiculous. I've got them on speed dial now. I call five times a day. When I wake up, when I go to work, during my lunch break at work, when I get home, and before I go to bed. They don't pick up anymore. But I still call!!!

Lighting shift.

ELSA: Dear Kamna, Apologies for my delay in response, but although I find your take very interesting, I cannot agree with your conclusions. We must not let ourselves off the hook with postmodern relativism. Perhaps I betray my aesthetic bias, but I do not think that that kind of thinking has gotten us anywhere, politically. If everything is de-centered, if climate change is so far beyond ourselves, there's no room for moral responsibility, is there? But there *is* moral responsibility. There *is* Amina. We must find ways to communicate across cultures—so that Americans can understand that this is not some abstract phenomenon. Climate change does have a face, and that face is Amina. We must not lose sight of her. I believe this very strongly.

ACTOR 1: The butterflies stayed in the curator's nest for over a month. They infested it. They ate all the time. They slept at odd hours. They drank the vodka in her freezer and made a mess of her sink. Then they took all her treasured pictures of Bangladesh and hibernated in their room.

Time passes.
Carla and Lewis compose the drawings.
The landscape shifts, collides, helps them.

ACTOR 2:

I'll build a castle in the mud

I'll sleep in mud and die in mud, be born in mud

be born from mud
from mud and shit and milk
I'll be a shit-smeared butterfly
like Carla and Lewis

The landscape is the drawing; the drawing is the landscape.

ACTOR 2:
I want to eat everything you eat
I want to sit in every chair you sit in
I want to wear every fabric you wear

They collapse.

LEWIS: I'm exhausted.

CARLA: Me too.

LEWIS: I can barely pick up my feet.

ELSA: Oh dear.

CARLA: This apartment is a sinkhole.

LEWIS: We're developing chronic fatigue here, Elsa. It's unhealthy.

CARLA: I think I have malaria.

LEWIS: Or leprosy.

ELSA: Maybe you should get out once and a while. I don't remember the last time you left. You've been working too hard.

LEWIS: Maybe she's right.

CARLA: You want to leave? With all our stuff here? Where are we going to go, India?

ELSA: Just to get some fresh air?

LEWIS: I think we have to. Look at your feet.

CARLA: Fine, but I'm taking my things with me.

LEWIS: Suit yourself. I'm leaving mine.

CARLA: You can't leave yours. You don't have anything.

LEWIS: I have duct tape, two spoons—

CARLA: THOSE SPOONS ARE MINE!

LEWIS: Fine, I have duct tape and a scarf!

ELSA: Maybe we could figure out a better place for you to work. Maybe we could move you to a studio space somewhere.

CARLA: Move us?

LEWIS: We don't want you to move us.

CARLA: We hate moving.

LEWIS: We just got here and now you want us to move?

ELSA: Okay! I'm just trying to help. You're difficult, you know?

CARLA: Of course we're difficult.

LEWIS: We're different from you.

ELSA: I realize that.

CARLA: Do you?

ELSA: I'm beginning to see it quite clearly, yes! I realize you have different skills from me, but I have skills, too! Excuse me for wanting a harmonious exchange of ideas!

LEWIS: Okay—okay!

ELSA: You know, you girls may be too young to understand, but you get to a certain age and you look around you and you say, my god, the world is a mess, and what have I been doing to clean it up? You get to a certain age and you want to stand in the corner of a park with a sign that says "The World is Ending"!

LEWIS: The corner of a park, really?

ELSA: It was a figure of speech!

CARLA: You're angry at us!

ELSA: No—no, I'm not angry.

LEWIS: Yes you are, you're very angry!

ELSA: NO, I'M NOT ANGRY, I'm just a little frustrated but I'll calm down, I will, just one second. I'm sorry for yelling at you. I didn't mean it.

CARLA: You're frustrated and angry. It's obvious. Is it because we yelled at the Scientist?

ELSA: What? No! I mean—well yes! Maybe just a little bit! I mean you've been here for a month and you lock yourselves in my guest room and you make my bathroom smell like nail polish and you won't even show me what you've been working on! I mean how am I supposed to do my work if you won't even show me what you're drawing?

CARLA: She wants to see the drawing.

LEWIS: Why didn't you say so?

ELSA: BECAUSE I RESPECT THE AUTONOMY OF THE ARTIST!

CARLA: Autonomy?

LEWIS: We live in your house.

CARLA: We drink your milk.

LEWIS: We eat your chips.

CARLA: We steal your Xanax.

LEWIS: We leave stray hairs on your sheets.

CARLA: We track toe jam on your carpeting.

LEWIS: We leave skin grease on your dishes.

CARLA: We shit in your toilet.

LEWIS: We cough up snot in your sink.

CARLA: You want to see our drawing?

LEWIS: Here it is.

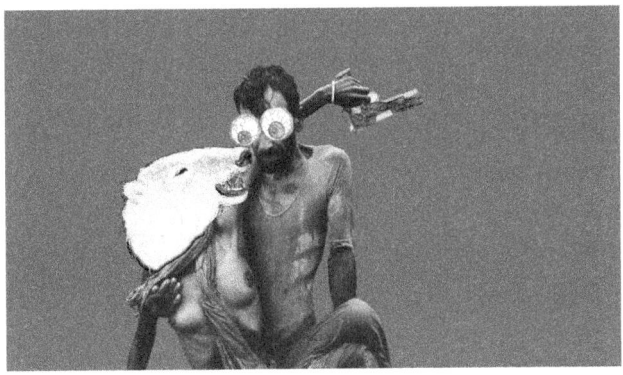

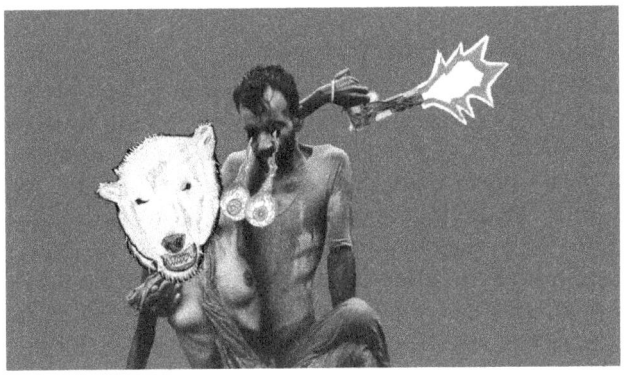

FIGURE 5.1 *Polar Bear and Bangladeshi man. Photo collage by Sunita Prasad*

They unveil the drawing.
Terrifying.

> ELSA: Oh—oh...

The landscape is the drawing; the drawing is the landscape.

> ACTOR 2:
> Here is the water
> Here is the mud
> Here are the mothballs
> Here is the mud
> Here are the rats, preening like birds
> Here is malaria
> Here is the mud
> Here are the trees who are illegal immigrants
> Here is rotting wood
> Dead fish
> Computer parts
> Here are the crocodiles
> Here is the mud
> Here is the mud
> Here is Amina
> Here is Amina
> Here is Amina
> Here is the mud

ACTOR 1: HUMANS! LEAVE YOUR HOMES!

KAMNA: Dear Ms. Turner, I must respectfully disagree with your assessment. There is no Amina. There can't be. There is no way to find her, and even if there were, there would be no way to understand climate change through her personal story. I have also begun questioning the other piece of the project—the dialogues over Skype. I worry that such conversations would be awkward and superficial, if not nonsensical. We must be wary of fetishizing the "real" Bangladeshi peasant. She may currently be the most visible of climate changes victims, but she is hardly the "face" of climate change. If there is a face of climate change, it's the fender of the American SUV. I would like to propose a different kind of dialogue—between artists. It so happens that I have the opportunity to travel to New York. My new husband's parents are giving us a gift of the trip for our honeymoon, and all I need is a letter of sponsorship from you so that I may obtain a visa.

Trees/illegal immigrants.

> **ACTOR 1:** There are different types of visas for traveling to the United States. Athletes, amateur and professional; au pair; Australian professional specialty; border crossing card (Mexico); business visitors, crewmembers; foreign military personnel
>
> **ACTOR 2:** I want to eat everything you eat
>
> **ACTOR 1:** HUMANS! LEAVE YOUR HOMES!
>
> **ACTOR 2:** I want to wear the same fabric as the one you're wearing
>
> I want to have the same taste in my mouth as the taste in your mouth
>
> I want to shape my muscles the way yours are shaped
>
> I want my folds of fat to mirror your folds of fat exactly
>
> I want my skin to be the same texture as your skin
>
> I want my hair to be the same texture, length, color, and style as your hair
>
> I want the contents of my gastro-intestinal tract to resemble absolutely the contents of yours

The trees/illegal immigrants fall down.

> **ELSA:** I don't understand.

Silence.

> **LEWIS:** Your apartment is very stimulating.
>
> **CARLA:** Stirring.
>
> **LEWIS:** Spawning, really.
>
> **ELSA:** My apartment? My apartment inspired THAT?
>
> **CARLA:** What else?
>
> **LEWIS:** We haven't left in a month.
>
> **ELSA:** But—all that black...
>
> **CARLA:** You mean the mud?
>
> **ELSA:** That's mud?
>
> **LEWIS:** Something mud-like, anyways. Nothing we know better than mud, at this point.
>
> **CARLA:** You know there's stuff growing in it? I think something's actually growing in between my toes!
>
> **ELSA:** What are you talking about?

LEWIS: The mud?

ELSA: But that's—that's just completely wrong!

Pause.
The weather changes.

LEWIS: Well if you don't like it...

ELSA: I don't! I don't like it at all!

Pause.

CARLA: Then I suppose this collaboration will just not work out.

ELSA: No. I suppose it won't.

LEWIS: Then I suppose you'll be wanting us to leave.

ELSA: Yes. I think that would be best.

CARLA: Right now?

ELSA: No, no, you don't have to leave right now. I'm sorry. I'm so sorry. It's my fault.

LEWIS: Of course it is.

ELSA: (*surprised*) You can stay until the end of the week.

CARLA: How generous.

LEWIS: Well we better go pack.

Lighting shift.

ELSA: Dear Ms. Banerjee, I apologize for my delayed response, I have been very busy. But I am afraid I am not in a position to sponsor your travel. I can't deal with another person here right now. Besides, the distance between our two countries is precisely what the project is supposed to address. And I do not want this to be a dialogue between artists. What I want are real people. Real people. Clearly you have not understood me at all. I am terribly sorry, but I do not think this collaboration will work out after all. I thank you for your time.

Carla and Lewis alone.

CARLA: I want to go back to Berlin.

LEWIS: I want to go back to our apartment.

CARLA: I want to sleep whenever I want and eat whatever I want and fuck whoever I want.

LEWIS: I want to roll around in whatever sidewalk scum I want and scream in public spaces in whatever language I want and ignore whatever white people I want!

CARLA: Clearly this is not a country you can do THAT in.

LEWIS: I'm sick of this mud.

CARLA: (*to audience*) And these GODDAMN FUCKING people who pretend they don't see it.

LEWIS: I don't want her fucking charity!

CARLA: I wouldn't stay in this shithole if she paid me!

LEWIS: And she BETTER pay us TOO!

CARLA: Elsa! ELSA!

Lights up on Elsa.

ELSA: I'm right here! What is it?

CARLA: We're getting out of here. Right NOW!

LEWIS: So you better pay us.

CARLA: You owe us at least three grand.

ELSA: Yes we do have to settle the question of remuneration—

LEWIS: Settle NOTHING!

CARLA: You made us live in your shitty dump for over a month.

LEWIS: This fucking hole is giving me malaria and I'm going to sue your ass right back to New Haven if you don't pay up!

CARLA: All our shit is ruined!

LEWIS: This mud is everywhere!

CARLA: And you better stop PRETENDING that it isn't!

ELSA: I KNOW, I KNOW IT'S EVERYWHERE, I keep calling and calling and NOBODY ANSWERS ME! I keep calling the building management and leaving 3, 4, 5 messages a week and does anybody listen? Does anybody CARE?

LEWIS: You obviously haven't been screaming loud enough, Elsa! You obviously haven't been screaming AS LOUD AS HUMANLY POSSIBLE!

ELSA: I HAVE, I have been screaming, how can I scream louder, they KNOW, they KNOW what's going on, and they just don't care, they just don't care that there's WATER in the WALLS!

CARLA: This isn't water, Elsa, it's MUD! It's MUD!

ELSA: I KNOW, I KNOW it's MUD, I just can't, I just can't do anything about it!

The scientist appears.

SCIENTIST: Will you keep it down, I'm trying to sleep!

ELSA: OH—Bronwyn! I'm sorry, I just—

CARLA: Don't apologize!

LEWIS: She's fucking mad!

CARLA: Her apartment is covered in MUD!

SCIENTIST: BIG GODDAMN DEAL!

ELSA: We were just! DISCUSSING! THE PROJECT!

SCIENTIST: Oh yes, the PROJECT. I forgot for a moments I was talking to a curator! I ask you seriously, could there be any job more obviously STUPID!

ELSA: I know it's stupid, you don't have to tell me it's stupid, I'm trying my hardest to make it NOT stupid and I don't need to be reminded that it's stupid by scientists with terrible haircuts!

SCIENTIST: You think this haircut's bad, huh? I AGREE! But what am I supposed to do about it?

ELSA: Get a better one!

SCIENTIST: Get a better job!

ELSA: Compost your garbage!

SCIENTIST: Stop taking planes!

ELSA: Don't yell at me!

SCIENTIST: Don't yell at ME!

ELSA: I'm trying my hardest!

SCIENTIST: So am I!

ELSA: It's very difficult!

SCIENTIST: Yes it is!

ELSA: We scream and we scream!

SCIENTIST: And nobody listens!

ELSA: What I really want to do is stand in the corner of a park with a sign that says

SCIENTIST: "THE WORLD IS ENDING!"

ELSA: But what good would that do!

SCIENTIST: None whatsoever! So I go back to the lab!

ELSA: So I think up a project!

SCIENTIST: AND I TRY TO FIND A WAY TO REPAIR THE DNA!

ELSA: And I try to build a work of art that might possibly change people's minds about something!

SCIENTIST: But then I realize!

ELSA: Nobody's going to change!

SCIENTIST: Nobody even cares!

ELSA: I'm just one woman! I'm just one white middle-class American woman, what power do I have in this world?

SCIENTIST: I have problems of my own to worry about!

ELSA: I'd like to one day own a house outside the city, maybe up the Hudson in one of those nice little towns and run a local gallery and grow my own vegetables!

SCIENTIST: Just because I like to take a bubble bath on Sunday evenings with a glass of red wine and go on the occasional tropical vacation with my gay best friend, does that make me the devil?

ELSA: No, no it doesn't! You have to take care of yourself, too!

LEWIS: That's right, Elsa!

CARLA: Take care of yourself!

LEWIS: Get this fucking mud leak fixed, Elsa!

CARLA: OH WAIT, YOU CAN'T!

LEWIS: You think if you IGNORE IT, it will GO AWAY?

CARLA: Don't you understand, the MUD IS RIGHT HERE!

An explosion, like a thunder clap and a dam breaking.
The mud lacerates the stage.
Blackout.
When the lights come up, Elsa is alone.

ELSA: I was sleeping, I was feeling warm, too warm, kicked off the cover, and half-woke, dreaming that someone, something had taken my sister away, a crocodile, into a dark marsh, where everything was dead and black, dead trees, no fruit, no leaves, teeth on the branches hung pain, and I say the crocodile is taking my sister, and I cry the salt of the marsh as he takes her all over, to the edge of the sea, the clouds are metal and the sun is a mango, and they're in a palace, red and purple fabrics and thunder and birds with yellow beaks and scales and rotting wood and red and purple feathers and green feathers and blue feathers which are now my sister's feathers, and in the corner of the palace is a little room, and in that little room is my little house, my house built with metal walls in the center of the palace on the edge of the black marsh and the crocodile who is my mother takes me in her arms and kisses me in the smooth point of my bare chest as moths fly out of her eyes and my sister is lying on a bed bleeding red, purple, green, blue, but I feel glad, I feel warm, too warm, kicked off the cover and then it happened, must have happened then, heard something, still dreaming, something like an animal maybe, walking by, a crocodile, except it was the wave, and I haven't woken up, but the wave is there and then I do wake up because it's in me and my chest breaks open and the wall of my house is my head, hard, wet, the mud, my chest, my shoulder, my bone, in the wall, which is gone, which is my chest, which is broken, which is the mud, in my mouth and throat and lungs, which thickens through my neck like a cake baked in my throat a mud cake. Bits of green plant and tails and fur from what and feathers red purple yellow beaks fish skin tiny rocks, a hard taste and ashy. I wake up to the mud cake which is my throat and the pain is horrible, the wall is my chest is broken, my bone is out of my body, brown in the water but it was just a second and the water disappears. And there is no wall and there is no house and there is no bed and there is no neck and there is no chest and there are no lungs, and there are no bones, the bones are the mud. I could have died but I could die anytime and if water killed us we would all die before birth. I didn't die. I woke up.

Program

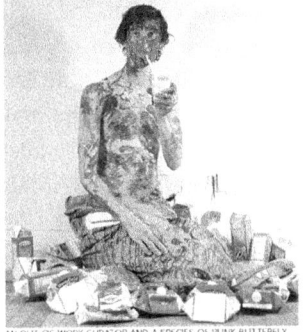

Cast

Nick Cregor and Daniel Squire: Actors
Libby King: Elsa Turner, a curator for the fine arts
Nicole D'Amico: Kamna Banerjee, A Bangladeshi mixed-media artist; Scientist
Meng Ai and Kim Rosen: Carla and Lewis, punk butterflies
The Company: Mud, Crocodiles, Malaria, Rats Preening Like Birds, Rotting Wood, A Subway Car that is also a Laundromat, Illegal Immigrants that are also Trees

Program

Bios

Una Chaudhuri (dramaturg) did her doctoral work at Columbia University. She has lectured internationally and published extensively on modern drama and performance theory. In recent years, Chaudhuri has been among the first scholars of performance to engage with the burgeoning field of Ecocriticism (which studies how environmental realities and discourses are reflected in literature, art, and the media) and the emerging field of Animal Studies (which studies the vast number of cultural animal practices human beings are involved in). This is her fourth collaboration with Fritz Ertl on their on-going research theatre project entitled "Theatre of Species."

Shonni Enelow (playwright) writes plays, performance pieces, cross-genre poetry, and cultural and literary criticism for both academic and popular audiences. Her recent solo performance piece, "My Dinner with Bernard Frechtman," premiered at the Invisible Dog art center in April 2010, and was the keynote performance at the American Literary Translators Association conference. Other plays have been performed in New York, San Francisco, Los Angeles, and at the Williamstown Theater Festival in Massachusetts. Her chapbook "Nietzsche is a Girl" was a Supermachine spirit gift; other poems have appeared in PomPom, Bird Dog, OnandOnScreen, and Try, among other publications. Her critical writing has appeared in Alef Magazine, Theater Topics, and *T: The New York Times* Style Magazine. She lives in Brooklyn. BFA, NYU.

Fritz Ertl (director) was co-founder of the award-winning BACA Downtown New Works Project in 1989. He has directed and produced new American plays by Mac Wellman, Paula Vogel, NeenaBeber, Erik Ehn, and Allan Havis at HOME, BACA Downtown, Hudson Guild Theater, Moving Target Theatre, and the Berkshire Theatre Festival. Most recently he directed Franz Xavier Kroetz's Mensch Meier at the Hudson Guild Theatre, and Severity's Mistress at New York University. Ertl has been working with the NYU Drama Department in New York for 18 years in both an administrative and creative capacity. His work as a teacher, director, and scholar has won him awards and recognition on two continents, as well as the deep respect of his peers and students.

Josh Hoglund's (director) recent directing and development work includes *My Dinner with Bernard Frechtman* (The Invisible Dog, 2010 ALTA Conference), *Taffy and Leon* (Ohio Theater 6th Fl. Series), *The Late*

Education of Sasha Wolff (CSV/Flanboyan Theater), all new plays written by Shonni Enelow. Other work includes *We The Emperor* (Montgomery St. Gardens; co-creator/performer), *This Time Tomorrow* (Duyrea Church workshop; performer), *International Love Cabaret* (Starr Space; performer), *MUD* (42nd Street Theater; director), *Far Away* (Robert Moss Theater; director), and *Karen Finley Presents: Think Hard, Darling Creatures* (The Culture Project; performer). Josh is a member of the Lincoln Center Theater Directors Lab. BFA, NYU.

Performers

Libby King is a member of the New York-based theatre company the TEAM. With the TEAM: Catalina in MISSION DRIFT (upcoming Lisbon's Culturegest, Edinburgh's Traverse Theatre), Carrie Campbell in ARCHITECTING (Co-Produced by the National Theatre of Scotland, Public Theatre, P.S.122, Barbican Centre, Traverse Theatre, Portugal's Publico Top Ten 2009), Sarah Springer in PARTICULARLY IN THE HEARTLAND (Best Female Performer Dublin Fringe, P.S.122, Walker Art Center, Time Out New York Top Ten 2007). Other credits: Astrov in Uncle Vanya (CSC), Girl in The Blind (CSC), Charley in Death of a Salesman (dir. Leon Ingulsrud), Willie in Tennessee Williams in Quarter Time (dir. John Dennis), Cordelia/Regan/ Goneril in LEARegardless (dir. Kristjan Thor) and Dottie in Killer Joe (MFA thesis role). Libby made her film debut in the movie *B.U.S.T.* (Special Jury Prize, Dallas Film Festival 2010). She is the educational coordinator for the TEAM's devising workshops and is currently on staff at NYU.

Nick Cregor is an actor, sound designer, video artist, and musician. Sound design credits include a Tischmainstage production of Dancing at Lughnasa, and an original play, *Movie Geek*, which won the Best Multimedia Fringe Festival Award in 2005. He has performed in the New York City area and toured across the U.S. as a musician under the name "72 south 1st." BFA, NYU.

Nicole D'Amico's favorite theater experiences include a national tour of *Charlotte's Web*, performing original work at the HERE Arts Center Summer Series and working with the physical comedy group Parallel Exit. Nicole is a member of Actor's Equity. She spends her days teaching music and movement to babies and toddlers in Manhattan. BFA, NYU.

Meng Ai's theater credits include *The Country Doctor* (Dir: Paul Lazar), *Hansel und Gretel* (Dir: David Neumann), *Full Circle* (Dir: Emma Griffin).

TV: *30 Rock* (NBC), *Royal Pains* (USA), *How to Make it in America* (HBO). Film: *Life During Wartime* (Dir: Todd Solondz), *Danger Island* (Dir: John Bruce). Meng can also be heard in China as the voice of the color commentator of WWE RAW and WWE SMACKDOWN. BFA, NYU.

Kim Rosen has previously appeared at The Public Theater, The Ensemble Studio Theater, 59E59, Galapagos Art Space, as the HERE Arts Center as Red Rose in Taylor Mac's Obie Award-winning epic *The Lily's Revenge!*, and as a member of the People's Sketch Association at The PIT, The Upright Citizens Brigade, The Tank, and most recently as part of the Boston Improv Festival. She is currently co-developing a web series with Alex Simons of *The New Yorker* entitled When Ladies Meet, roughly inspired by the Joan Crawford film of the same name. BFA, NYU.

Daniel Squire was born in Halifax, UK. He studied dance at White Lodge and at the Rambert School, concurrent with working as a percussionist in several semi-professional orchestras in Yorkshire and London. He then worked as a dancer with Michael Clark and Matthew Hawkins, as well as appearing as Tadzio in Britten's *Death In Venice* at Glyndebourne. After moving to the Big Apple, Daniel worked for many years with Merce Cunningham, performing around the world in theatres including PalaisGarnier&Théâtre de la Ville (Paris); Staatsoperunter den Linden & Schiller Theater (Berlin); the Roundhouse, the Barbican, Tate Modern (London); Kennedy Center (DC); New York State Theater, Brooklyn Academy of Music, the Rose Theater & City Center (NYC); with musicians and music groups including Radiohead, John Paul Jones, SigurRós, TakehisaKosugi & Sonic Youth. He continues to lives in Manhattan and works as a dance teacher (for internationally acclaimed dance companies and at top universities), a videographer, film-maker, editor, and actor.

Designers

Felix Ciprián (costumes) is a sculptor, costume designer, performer, living in New York City. Born in Miami, Florida, Felix grew up both in the Dominican Republic and New York City. He studied Theater at NYU's Tisch School of the Arts, graduating in 2005. While at Tisch he had the opportunity to study Irish culture and cinema in Dublin, Ireland in association with Trinity University. Upon graduating, Felix became involved in costume design. His work has been featured at the Theater for the New City, Center Stage, 45 Street Theater, Flamboyan CSV Theater, Natives Theater, Linhart

Theater, and La Mama. Since 2006, Felix has begun investigating the drama and narratives in discarded objects. He sifts through the wreckage to form emotionally laden sculptures which have an odd, and even quirky presence. His work has been featured in the Deitch Art Parade, Real Art Ways, Chashama, Studio 717, Art Gotham, Ephemorptera Art Space and the Stephen Savage Gallery. He currently lives in Brooklyn, NY. Please visit felixanddexter.com. *Carla and Lewis* marks David Herman's third exploit alongside many on the C+L team, having previously designed sound for *The Late Education of Sasha Wolff* and *My Dinner with Bernard Frechtman* (both written by ShonniEnnelow and directed by Josh Hoglund). By day, David is an engineer and sound designer for WNYC's Freakonomics Radio, and serves as one half of the partnership behind Launderette Recordings, a music label focused on preserving nearly lost historical sounds.

Sayaka Nagata (set and objects) was born and raised in Japan. Sayaka has been breathing New York air for the past 14 years. Her past collaborators include Gerrit Turner (*Othello* by William Shakespeare), Ted Walter (*Not, Not, Not, Not, Not Enough Oxygen* by Caryl Churchill, *The Investigation of the Murder in El Salvador* by Charles Mee) and Laura Shiffrin (*The Maids* by Jean Genet) among others. Long-time collaborator to Josh Hoglund, she has worked on projects such as *The Late Education of Sasha Wolff* by Shonni Enelow, *Far Away* by Caryl Churchill, *Sincerity Forever* by Mac Wellman with him. This is her third project as a set designer for Josh. She also makes paintings. BFA, NYU. Please visit sayakanagata.com

Sunita Prasad (video and visuals) makes films, videos, performances, and photographs. This is her third project with Josh, Shonni, and the other designers of *Carla and Lewis*, for which she is very grateful. Her solo work has shown at festivals and galleries nationally, though her most recent project, Imaginary Islands, was happily supported and exhibited by her home borough of Brooklyn NY. For more information, please visit sunitaprasad.net.

Dramaturg's note: The Weather Changes

Una Chaudhuri

For the first time in human history, the weather is about us. So now, when the weather changes, shouldn't we? Aren't we? Haven't we already? What is it like to be the climate change generation—the ones who have to admit that it's happening now, not later, and it's happening to us, not just to others elsewhere? What emotions—beyond paralyzing guilt and despair—, and what genres—besides slide-shows and documentaries—speak to this ecological reality?

This project, like others we've worked on previously, unfolded as a "research theatre" process that began with a couple of texts, and a question.

The question was: how might current theories of biology and ecology be brought into the time and space of live performance? How might the theatre contribute to the recognition that *Homo sapiens* is one species among many, and one that is in need of new ideas and new lifeways in the face of catastrophic climate change?

The texts were a novel—*An Incident in the Life of a Landscape Painter*, by Argentine author Cesar Aira—and a brief theoretical essay, entitled "Queer Ecology," by Timothy Morton, author of a recent and important ecocritical book, *Ecology Without Nature*. (Other texts that influenced our thinking as the process went on were *No Future: Queer Theory and the Death Drive*, by Lee Edelman; *A Natural History of Sex: The Ecology and Evolution of Mating Behavior*, by Adrian Forsyth; and *Darwin's Dangerous Idea*, by Daniel Dennett).

Aira's novel tells the story of a landscape painter who is struck by lightning and plunged into a state of agonizing insight. Continuing to record the landscapes around him despite his debilitating condition, he achieves a kind of fantastic literalization of the visionary natural history of his teacher, Alexander Von Humboldt. The result is something like living out a new relationship to nature: in effect, to *becoming landscape*. This furious "becoming" entails a violent merging of the artist's body with the artwork and its subject, a material fusing—on the *molecular* level—that also reminded us of Artaud's dream of a life-filled theatre.

Morton's essay links the anti-essentialist, performative impulse of queer theory to an emergent anti-essentialist, fully *relational* yet non-systematic ecology, based on the growing acknowledgment of the profound anti-essentialism of Darwinian evolutionary theory. In this view, evolution is neither linear, nor progressive, nor purposive. Rather, it is digressive and transgressive. Morton describes this "queer ecology" as one where boundaries are "blurr[ed] and confound[ed] at practically any level: between species, between the living and the nonliving, between organism and environment." In combination with Donna Haraway's theorization of *Homo sapiens* as a "companion species," queer ecology offers an alternative to the environmental tradition based on capital-N Nature, "out-there" Nature, and to its synoptic visions and holistic fantasies. In place of the pristine wilderness and sacred "biotic community" of that tradition, queer ecology imagines a "coming collectivity," ever emergent, and ever evanescent.

Carla and Lewis is imagined both as a play and as an algorithm for producing a theatrical landscape hospitable to versions of this coming collectivity. We want to use the insights and marvels and conventions and tropes of postmodern theater to push towards something new. All the elements of theatre are invited to or pushed to partake in evolutionary logic: time, space, objects, persons, events, and gestures adopt a theatrical version of descent with modification: repetition and revision. The resulting landscape—ever emergent and ever evanescent—feels like the right setting for the story we want to tell. Like Aira's, ours is also a story of furious becomings, unique intimacies, sudden shifts, endless accumulations.

Carla and Lewis borrows its central figure for the contemporary discourse on ecological catastrophe from a classic of queer theory: Eve Sedgewick's *Epistemology of the Closet*. The skeletons in the closet now are no longer linked to sexual identity but to species identity. The Climate Change Closet is stuffed with the denials and anxieties of a culture that has been at war with the non-human world for centuries. We can see now that Artaud sensed its presence, and tried to resist its epistemologies: "All our ideas about life must be revised in a period when nothing any longer adheres to life; it is this painful cleavage which is responsible for the revenge of *things*." Artaud seems to echo the old man in Ibsen's visionary proto-ecodrama, *The Wild Duck*, whose explanation for tragic loss is "The woods take revenge!" In *Carla and Lewis*, the mud of Bangladesh takes revenge, and its action is no longer operating at a distance. Just as contemporary artists seek actual encounters and meaningful proximities, and just as the forces of global capital compel transnational flows of images, things, and substances, climate change takes its revenge—or makes its points—*in person*, face to face, body with bodies. *Carla and Lewis* is an attempt to link the resources of live performance—the shared experience of living bodies inhabiting the same time and space—to the pervasive, intrusive, and above all *intimate* modalities of ecology understood in all its queerness.

Bibliography

Aira, César. *An Episode in the Life of a Landscape Painter*. Translated by Chris Andrews. New York: New Directions, 2006.

Arons, Wendy and Theresa J. May. *Readings in Performance and Ecology*. Basingstoke: Palgrave Macmillan, 2012.

Artaud, Antonin. *The Spurt of Blood*. Translated by Ruby Cohn. *Theater of the Avant-Garde 1890–1950*. Edited by Bert Cardullo and Robert Knopf. New Haven: Yale University Press, 2001.

——. *The Theater and Its Double*. Translated by Mary Caroline Richards. New York: Grove Press, 1958.

Baker, Steven. *The Postmodern Animal*. London: Reaktion Books, 2000.

Bennett, Jane. *Vibrant Matter*. Durham, NC: Duke University Press, 2010.

Berger, John. "Why Look at Animals?" In *About Looking*. New York: Pantheon, 1980.

Bolton, Andrew. *WILD: Fashion Untamed*. With contributions by Shannon Bell-Price and Elyssa Da Cruz. New York and New Haven: Metropolitan Museum of Art and Yale UP, 2004.

Brewster McCloud. Dir. Robert Altman. Perf. Bud Cort, Shelley Duvall, Sally Kellerman. Lion's Gate, 1970. Film.

Chakrabarty, Dipesh. "Postcolonial Studies and the Challenge of Climate Change." *New Literary History* 43, no. 1 (Winter 2012).

Chaudhuri, Una. "There Must Be a Lot of Fish in That Lake: Towards an Ecological Theatre," *Theater* 25, no. 1: (1994): 23–31.

———. "Animal Acts for Changing Times." *American Theater* October 2004: 36–39.
———. "Animal Geographies: Zooësis and the Space of Modern Drama." *Modern Drama* 46, no. 4 (Winter 2003): 646–62.
Chaudhuri, Una and Shonni Enelow. "Animalizing Performance, Becoming-Theater: Inside the Animal Project at NYU," *Theater Topics* 16, no. 1 (March 2006): 1–17.
Cless, Downing. *Ecology and Environment in European Drama*. London: Routledge, 2010.
Coetzee, J[ohn] M[ichael]. *The Lives of Animals*. Princeton: Princeton University Press, 2001.
Deleuze, Gilles, and Felix Guattari. "1730: Becoming-Intense, Becoming-Animal, Becoming-Imperceptible...." In *A Thousand Plateaus: Capitalism and Schizophrenia*. Translated by Brian Massumi. Minneapolis: University of Minnesota Press, 1987.
Edelman, Lee. *No Future: Queer Theory and the Death Drive*. Durham, NC: Duke University Press, 2004.
Engel, Marian. *Bear*. 1976. Boston: Nonpareil Books, 2004.
Genet, Jean. *The Declared Enemy: Texts and Interviews*. Edited by Albert Dichy. Translated by Jeff Fort. Stanford, CA: Stanford University Press, 2004.
Haraway, Donna. *The Companion Species Manifesto: Dogs, People, and Significant Otherness*. Chicago: Prickly Paradigm Press, 2003.
Kershaw, Baz. *Theatre Ecology: Environments and Performance Events*. Cambridge: Cambridge University Press, 2007.
Keyes, Ken. *The Hundredth Monkey*. St. Mary, KY: Vision Books, 1981.
Koestenbaum, Wayne. "Situation Hacker: Wayne Koestenbaum on the Art of Ryan Trecartin." *ArtForum International*, June 22, 2009.
Latour, Bruno. "Love Your Monsters: Why We Must Care for Our Technologies as We Do for Our Children." In *Love Your Monsters: Postenvironmentalism and the Anthropocene*. Edited by Michael Shellenberger and Ted Nordhaus (Breakthrough Institute, 2011). Ebook.
———. *Reassembling the Social: An Introduction to Actor-Network Theory*. London and New York: Oxford University Press, 2007.
Mortimer-Sandilands, Catriona and Bruce Erickson, eds. *Queer Ecologies: Sex, Nature, Politics, Desire*. Bloomington, IN: Indiana University Press, 2010.
Morton, Tim. "Guest Column: Queer Ecology." *PMLA* 125 no. 2 (2010): 273–282.

Müller, Heiner. *Hamletmachine and Other Texts for the Stage.* Edited and translated by Carl Weber. New York: Performing Arts Journal Publications, 1984. Print.

Nixon, Rob. *Slow Violence and the Environmentalism of the Poor* (Cambridge: Harvard University Press, 2011).

Read, Alan, ed. *On Animals.* Spec. issue of *Performance Research* 5 no. 2 (Summer 2000).

Shay, Arthur. *Animals.* Urbana: University of Illinois Press, 2002.

Shellenberger, Michael and Ted Nordhaus. "Evolve: The Case for Modernization as the Road to Salvation." *Love Your Monsters: Postenvironmentalism and the Anthropocene.* Edited by Shellenberger and Nordhaus (Breakthrough Institute, 2011). Ebook.

Spivak, Gayatri Chakravorty. *A Critique of Postcolonial Reason: Towards a History of the Vanishing Present.* Cambridge, MA: Harvard University Press, 1999.

Stein, Gertrude. *Last Operas and Plays.* Edited by Carl Van Vechten. Baltimore, MD: Johns Hopkins University Press, 1995.

Taylor, Jesse Oak. "The Novel as Climate Model: Realism and the Greenhouse Effect in *Bleakhouse.*" *Novel: A Forum on Fiction* 46 no. 1 (2013).

Urpeth, James. "Animal Becomings." In *Animal Philosophy: Ethics and Identity.* Edited by Peter Atterton and Matthew Calarco. London: Continuum, 2004.

Wolfe, Cary, ed. *Zoontologies: The Question of the Animal.* Minneapolis: University of Minnesota Press, 2003.

Index

actant, 34–36, 38, 76
Aira, César, 35–36, 50–54, 75
An Episode in the Life of a Landscape Painter, 35–36, 50–54
animal studies, 2, 11, 20
anthropocene, 25, 27, 33
Artaud, Antonin, 27, 34–35, 36. See also eco-cruelty

Bangladesh, 32, 36, 66–67, 77. See also Dhaka
becoming, 11, 13–15, 17, 18, 33, 36, 38, 64, 70, 75, 76, 77
becoming-intense, becoming-animal, becoming-imperceptible, 11, 17
Bennett, Jane, 30, 34
Bentley, Eric, 5
The Playwright as Thinker, 5
Berger, John, 11
"Why Look at Animals?",
Brecht, Bertolt, 8, 43, 76

Chakrabarty, Dipesh, 24–28
climate change, 2, 23–28, 31–32, 33, 36, 38, 45, 49, 65, 66–67, 76, 77, 78
climate change refugee, 2, 32, 65, 79
representation of, 23
Coetzee, J.M., 11
collaboration, 42–44, 47–60, 68, 76

consumer capitalism, 5, 7

Darwin, Charles, 30, 43, 44, 46, 64
Deleuze, Gilles and Felix Guattari, 11, 13–14, 18, 31, 36, 43–46, 64, 75
A Thousand Plateaus, 11
deterritorialization, 15, 18, 36, 75
Dhaka, 69. See also Bangladesh
documentary theatre, 5, 8
Donis, Alex, 9
doubled human, 28, 33

eco-cruelty, 27, 36
ecotheatre, 2, 28–30
Edelman, Lee, 31, 46, 47, 49
etude, 4, 16–20, 43–50, 52–53, 56, 57, 60
evolution, 30, 39, 45, 48, 51, 52, 57, 60, 63, 64, 78

GIF animations, 69–70
globalization, 2, 3, 5, 11
ground zero, 3, 77, 88

Hamlet, 13, 15, 18
Haraway, Donna, 11, 13, 18
The Companion Species Manifesto: Dogs, People, and Significant Otherness, 11, 18
herd, 17, 31, 33

home, concept of, 36
Humboldt, Alexander von, 35, 51, 56, 60
Husain an Nim Nim, 9

imperative, 4, 43–45, 51–53, 58, 59
improvisation, 13, 42, 58, 72
interactive theatre, 77
intimacy, 26, 30, 31, 46–47, 48, 60, 63, 64, 77
Iraq, 7–9

Joint Stock theatre company, 42, 53

landscape, 9, 10, 30–33, 35–36, 38–39, 50–53, 63–64, 66, 71–72, 74, 75–78
  becoming-landscape, 33, 35–36, 38, 75–76. *See also* becoming
landscape of catastrophe, 8, 49, 56, 61
Latour, Bruno, 26, 34, 76
literalism, 28, 29

Maldives underwater cabinet meeting, 28. *See also* Mohamed Nasheed
materialism, 2, 15, 26, 75–76
Mohamed Nasheed, 24
molecularization, 14, 64, 65
Morton, Timothy, 29–31, 46–47
Müller, Heiner, 76

neo-liberalism, 5, 6
New York City, 3, 19, 69
Nixon, Rob, 24
No Future, 35, 46, 47, 49. *See also* Edelman, Lee

posthumanism, 2
punk, 30–31, 65, 67–69, 74–78

queer ecology, 8, 26, 30–31, 35, 46–47. *See also* Morton, Timothy

research theatre, 1–5, 7, 9–11, 20, 41–44, 53, 74, 78

September 11, 2001, 3, 4, 7
Sgt. Mkesha Clayton, 8
Shay, Art, 16
Shellenberger, Michael and Ted Nordhaus, 26
silence, 3, 4. *See also* imperative
slow violence, 24. *See also* Nixon, Rob
Species Theatre, 45, 49, 50
spectatorship, 76
Spivak, Gayatri Chakravorty, 74
Stein, Gertrude, 76
subaltern, 32, 36, 74, 78. *See also* Spivak, Gayatri Chakravorty
suicide environmentalism, 6
  the Church of Euthanasia, 6

The Animal Project, 11, 13, 18, 20
The Invisible Dog, 58, 75
The Queerak Project, 7, 8, 10, 11
The Resistance Project, 3, 5, 11
Theatre of Species, 11

utopian counterfactual, 9

war, 2, 8, 9, 10
  cold war, 3
  Iraq war, 7, 8

GPSR Compliance

The European Union's (EU) General Product Safety Regulation (GPSR) is a set of rules that requires consumer products to be safe and our obligations to ensure this.

If you have any concerns about our products, you can contact us on

ProductSafety@springernature.com

In case Publisher is established outside the EU, the EU authorized representative is:

Springer Nature Customer Service Center GmbH
Europaplatz 3
69115 Heidelberg, Germany

www.ingramcontent.com/pod-product-compliance
Lightning Source LLC
LaVergne TN
LVHW041205250326
834689LV00002BA/23